Business Guide

"Business Guides on the Go" presents cutting-edge insights from practice on particular topics within the fields of business, management, and finance. Written by practitioners and experts in a concise and accessible form the series provides professionals with a general understanding and a first practical approach to latest developments in business strategy, leadership, operations, HR management, innovation and technology management, marketing or digitalization. Students of business administration or management will also benefit from these practical guides for their future occupation/careers.
These Guides suit the needs of today's fast reader.

More information about this series at
http://www.springer.com/series/16836

Martin Lempp • Patrick Siegfried

Automotive Disruption and the Urban Mobility Revolution

Rethinking the Business Model 2030

Springer

Martin Lempp 📴
International Business
ISM International School of
Management GmbH
Oberursel, Germany

Patrick Siegfried 📴
International Management
ISM International School of
Management GmbH
Frankfurt am Main, Germany

ISSN 2731-4758 ISSN 2731-4766 (electronic)
Business Guides on the Go
ISBN 978-3-030-90038-0 ISBN 978-3-030-90036-6 (eBook)
https://doi.org/10.1007/978-3-030-90036-6

© The Editor(s) (if applicable) and The Author(s), under exclusive licence to Springer Nature Switzerland AG 2022

This work is subject to copyright. All rights are solely and exclusively licensed by the Publisher, whether the whole or part of the material is concerned, specifically the rights of translation, reprinting, reuse of illustrations, recitation, broadcasting, reproduction on microfilms or in any other physical way, and transmission or information storage and retrieval, electronic adaptation, computer software, or by similar or dissimilar methodology now known or hereafter developed.

The use of general descriptive names, registered names, trademarks, service marks, etc. in this publication does not imply, even in the absence of a specific statement, that such names are exempt from the relevant protective laws and regulations and therefore free for general use.

The publisher, the authors and the editors are safe to assume that the advice and information in this book are believed to be true and accurate at the date of publication. Neither the publisher nor the authors or the editors give a warranty, expressed or implied, with respect to the material contained herein or for any errors or omissions that may have been made. The publisher remains neutral with regard to jurisdictional claims in published maps and institutional affiliations.

This Springer imprint is published by the registered company Springer Nature Switzerland AG.
The registered company address is: Gewerbestrasse 11, 6330 Cham, Switzerland

Preface

From an abstracting perspective, the urban infrastructure offers quite apparent analogies to a vivid organism circulating matter throughout connecting paths to all destinations in the capillary systems. A putatively chaotic process of parallel and consecutive movements sustains the functional capability of the organism. In our allegory, the complex road and railway network that pervades our metropoles like a vascular system preserves the mobility for citizens and commodities. However, evolving demographics and the incessant trend towards urbanization exert pressure on the capacity of the prevailing mobility concept. Technological progress offers new possibilities to arrange the chaos and increase the efficiency of our infrastructure. Automotive manufacturers need to rethink the durability of their established business models that are susceptible to evolving demand patterns induced by advances in connectivity and autonomous driving. The digitalization and internet-based interlinkage of people accelerates the diffusion of versatile innovations and leads to a more distinguished differentiation of mobility requirements. The demand for flexible mobility, individual lifestyles, budget restrictions, or varying user experiences sharpens the existing and emerging consumer profiles. The renewal of the business model 2030 is not only limited to OEMs but also embraces tier-n-suppliers and all stakeholders that seek to exploit the value-driven service potentials arising from the new mobility concepts. The development of intra-infrastructure communication, car-to-car integration, and service offers that enhance the value-of-time is conducted and concurrently affects an increasing number of automotive stakeholders.

Times of disruption are always accompanied by uncertainties that may either inspire or paralyze corporate progress. Which technological trend offers the most promising growth? Which allies and antagonists are gathering within the market? And how do companies manage to benefit from shifting key markets and sources of revenue? Customer surveys and studies of incremental demographic changes contour the prospective business environment that companies will have to operate in. The acceptance of technological innovations, willingness-to-pay for additional services, and the overall market size determine how much effort the automotive industry is willing to make to access new customer groups and product niches.

The prevailing corporate structures and vertical alignment of the value chains are challenged by the need to bundle innovative capacities, cost-efficient production as well as sensitivity and reactivity to changes in consumption behavior. Blurring boundaries between manufacturers, tier-n-suppliers, and IT-service providers require new, mutually beneficial methods of collaboration and co-operation and may shift market dominance from pure manufacturers to globally operating, data-driven companies that provide the basis for the interconnectivity and enhanced customer services. The direction and size of profit flow emanating from the customers throughout the product life cycle are increasingly rerouted and guided towards a growing number of recipients. The lion's share of turnover might prospectively not be made by the physical basic product but with additional services provided by the imbedded infrastructure such entertainment features, real-time navigation, data-based car-to-car connectivity, car and workplace interlinkage, or remote onboard diagnostic. The ability of the vehicle to digitally communicate with all entities providing these supplemental services represents the catalyzer for further innovations that will shape the business models in the forthcoming decades even beyond the business model 2030. Despite a lot of research conducted in the automotive field, studies published by specialized consultancies, or interviews given by leading managers of the industry, there is still no integrated perspective of the automotive market for the next

10–15 years. Both customers and producers are shaping the market simultaneously and the first steps of the mobility revolution have already been taken. The automotive companies have to strike new paths to participate in this journey.

After introducing the reader to the subject of this research work by illustrating the objectives and methodological structure, all relevant characteristics of the automotive industry are depicted including prevailing business models of OEMs and tier-n automotive suppliers, the competitive environment they are imbedded in as well as socio-economic changes affecting future market conditions. Subsequently, elements of the automotive disruption are presented—this research work focuses on connectivity and autonomous driving—that enable the provision of novel urban mobility concepts and offer a new source for additional services accompanying the user. The quantitative analysis of the customer perspective will, based on surveys conducted within the prospective customer group, help to determine the future market size and give a comprehensive insight into consumer behavior. Based on the previous analysis, potential automotive business models 2030 are portrayed whereof future development scenarios for automotive manufacturers, smart city models, and differing penetration scenarios are deducted. Given these scenarios, transformation strategies are depicted that might change the horizontal and vertical structure as well as the revenue distribution of the overall market and the companies within. To conclude the research work, challenges and key actions that shape the automotive sector even beyond 2030 as well as knock-on effects across different industries arising from the technological and economic changes in the automotive market are projected.

Oberursel, Germany
Frankfurt am Main, Germany

Martin Lempp
Patrick Siegfried

Contents

1 **Introduction to Automotive Disruption and the Urban Mobility Revolution**   1
    1.1   Problem Definition   1
    1.2   Hyporesearch Work   3
    1.3   Methodological Structure   4
    References   5

2 **Characterization of the Automotive Industry**   7
    2.1   Traditional Business Models of OEMs and Market Profile   7
    2.2   Gearing Up the OEM–Supplier Interface   11
    2.3   Key Trends Affecting the Mobility Market   19
    References   20

3 **Elements of the Automotive Disruption**   25
    3.1   Disruptive Mobility and IT Trends   25
        3.1.1   Connectivity and the Internet of Things   25
        3.1.2   A Foresight into the Era of Autonomous Driving   27
        3.1.3   Urban Mobility Concepts and the Electrification of the Car   30
    3.2   Digital Transformation   35
        3.2.1   Data—The Emergence of a New Currency   35

		3.2.2	Digital Value Chain and Value Creation Models	37
		3.2.3	Creation of the Digital Ecosystem	44
	References			49

4 Urban Mobility Revolution: A Quantitative Analysis — 55
- 4.1 Predicted Customer Demand Patterns — 55
- 4.2 Value of Time and Willingness-to-Pay for Digital Services — 74
- 4.3 Valuation of the Future Market Volume — 80
- References — 90

5 Business Model 2030: A Metamorphosis of the Automotive Landscape — 95
- 5.1 Development Scenarios — 95
  - 5.1.1 Creation of the Smart City — 95
  - 5.1.2 OEM Scenarios: Forms of Repositioning — 108
  - 5.1.3 Mobility Scenarios: Extension of the Mobility Landscape — 110
- 5.2 A New Appreciation of the Value Chain — 115
  - 5.2.1 Value Chain Integration: Traditional Versus Digital — 115
  - 5.2.2 The Significance of Partnerships and Strategic Alliances — 117
- 5.3 Defining Use Cases in the Disrupted Automotive Market — 124
  - 5.3.1 Determining the Accessibility of new Revenue Sources — 124
  - 5.3.2 Redistribution and Monetization of Revenue Pools — 128
  - 5.3.3 Excursus: Upheaval in the Insurance Industry — 135
- References — 138

6 Conclusion to Automotive Disruption and the Urban Mobility Revolution — 147
- 6.1 Outlook 2030 — 147

	6.2	Limitations of the Research Work	149
	References		150
7	**Prospects: A Look Beyond**		153
	7.1	Future Aspects	153
	References		155

Abbreviations

AI	Artificial Intelligence
BOT	Build-Operate-Transfer
BRIC	Coalition of emergent economies (Brazil, Russia, India, China)
CAGR	Compound Annual Growth Rate
CEM	Customer Experience Management
CRM	Customer Relationship Management
E-Hailing	Electronic Hailing (private taxi services)
ERP	Enterprise Resource Planning
EV	Electric Vehicles
GDPR	General Data Protection Regulation (Datenschutz-Grundverordnung DSGVO)
HaS	Hub-and-Spoke
ICE	Internal Combustion Engine
ICT	Information & Communication Technology
IP	Intellectual Property
IT	Information Technology
JIS	Just-in-Sequence
JIT	Just-in-Time
kWh	Kilowatt hour
Lidar	Light detection and ranging
MaaS	Mobility-as-a-Service
NHTSA	National Highway Traffic Safety Administration
OEM	Original Equipment Manufacturer

P2P	Peer-to-Peer
PayD	Pay-as-you-Drive
ROI	Return on Investment
R&D	Research & Development
SAE	Society of Automotive Engineers
SDV	Self-Driving Vehicle
SUV	Sport Utility Vehicle
TCO	Total Costs of Ownership
UbI	Usage-based Insurance
V2I	Vehicle-to-Infrastructure
V2V	Vehicle-to-Vehicle
V2X	Vehicle-to-Everything
VDA	Verband der Automobilindustrie/Association of the Automotive Industry
VoT	Value of Time
WLTP	Worldwide harmonized Light vehicle Test Cycle

List of Figures

Fig. 2.1	Short-term and long-term trends in the automotive industry. Roland Berger and Lazard (2016), S.17	10
Fig. 2.2	Key trends affecting the OEM–supplier relationship. The Boston Consulting Group (2004), S.12	12
Fig. 2.3	Six levers to optimize the OEM–supplier interface. The Boston Consulting Group (2004), S.20	17
Fig. 2.4	Collocations contribute to an effective cooperation between purchasing and R&D. The Boston Consulting Group (2004), S.25	18
Fig. 3.1	Roadmap towards fully autonomous driving. Wyman, Oliver (2017), S.49	28
Fig. 3.2	SDV benefits for individuals and society. The Boston Consulting Group/ World Economic Forum (2016) S.14	29
Fig. 3.3	The increased use of car data will unlock new customer benefits in four areas. McKinsey & Company II, (2016), S.8	30
Fig. 3.4	Global trends triggering change in the mobility industry. McKinsey & Company III, (2016), S.10	31
Fig. 3.5	New mobility services offer transportation alternatives. McKinsey & Company II, (2015), S.13	34
Fig. 3.6	Car data users/contributors and use case examples. McKinsey & Company II (2016) S.5.	37
Fig. 3.7	Potential supply chain automation applications (smart planning). PA Consulting, (2018), S.30	39

xv

Fig. 3.8	Potential supply chain automation applications (smart sourcing). cf. PA Consulting Group; The Consumer Goods Forum (2018) AI and Robotics automation in consumer-driven supply chains S.30	40
Fig. 3.9	Potential supply chain automation applications (smart manufacturing). PA Consulting, (2018), S.30/31	41
Fig. 3.10	The restructuring of traditional value chains. A.T. Kearney (2016) How Automakers can survive the Self-Driving Era S.13	43
Fig. 3.11	Four distinct types of ecosystems. Author's own representation based on BearingPoint/IIHD Institute, (2017), S.8	45
Fig. 3.12	Network effects exemplified with Uber. Author's own representation based on BearingPoint/IIHD Institute, (2017), S.14	47
Fig. 4.1	Demographic characteristics of the survey group. Author's own representation	56
Fig. 4.2	Digital mobility interest levels and the expectations for future mobility scenarios	58
Fig. 4.3	Would you use an autonomous vehicle? Source: Author's own representation	60
Fig. 4.4	In your opinion, which reasons argue against the usage of autonomous vehicles? Source: Author's own representation	61
Fig. 4.5	In your opinion, which reasons argue for the usage of autonomous vehicles? Source: Author's own representation	62
Fig. 4.6	Which purchase criteria do you consider the most important? Source: Author's own representation	63
Fig. 4.7	Customer segments that determine the pace of market penetration (McKinsey & Company, 2014, S.12)	66
Fig. 4.8	Participants' willingness to provide private and vehicle-related data. Source: Author's own representation	69
Fig. 4.9	Macro-categories of data with different levels of perceived privacy sensitivity (McKinsey & Company II, 2016, S.16)	70
Fig. 4.10	In your opinion, which of these companies will lead in the development of autonomous vehicles? Source: Author's own representation	72
Fig. 4.11	How often do you use the following means of transport? Source: Author's own representation	73
Fig. 4.12	Society of Automotive Engineers (SAE) Vehicle Automation Levels. National Highway Traffic Safety Administration (NHTSA) (n.a.)	75

List of Figures   xvii

Fig. 4.13	What would use your disposable time for when driving in an autonomous vehicle? Source: Author's own representation	77
Fig. 4.14	Forecast of new car sales in the EU/US and China (PwC's Strategy&, 2017, S.8)	82
Fig. 4.15	Total market value split of vehicle modules and future development (Oliver Wyman, 2017, S.33)	83
Fig. 4.16	Installed vehicle bases from 2017 to 2030 (in millions) (PwC's Strategy&, 2017, S.9)	84
Fig. 4.17	Forecasted mobility development in the EU/US and China (PwC's Strategy&, 2017, S.18)	85
Fig. 4.18	Estimated Mobility-as-a-Service market size developments in the EU/US and China (PwC's Strategy&, 2017, S.19)	87
Fig. 4.19	Industry distribution of household mobility expenditure in shared autonomous scenarios, Distribution of Western premium household spend (in US$) (PwC's Strategy&, 2017, S.26)	88
Fig. 4.20	Global self-driving minutes (long-distance commuting only) (A.T. Kearney, 2016, S.9)	89
Fig. 5.1	A framework for understanding urban mobility. McKinsey & Company II (2015) S.4	96
Fig. 5.2	Annual cost of mobility in San Francisco Bay Area (in US$). McKinsey & Company II (2015) S.9	99
Fig. 5.3	Strategic alliances in the field of V2X communication. A.T. Kearney (2016) S.22	101
Fig. 5.4	Connectivity features will evolve along four dimensions. A.T. Kearney (2016) S.31	102
Fig. 5.5	Municipal Traffic Systems will help control and steer SDVs. The Boston Consulting Group/World Economic Forum (2016) S.16	107
Fig. 5.6	The impact of car sharing, urbanization, and macroeconomics. McKinsey & Company I (2016) S.10	111
Fig. 5.7	Fully autonomous vehicle share of new vehicle market. McKinsey & Company I (2016) S.11	112
Fig. 5.8	Increasing complexity of the competitive landscape. McKinsey & Company I (2016) S.13	116
Fig. 5.9	The supply side of connected cars: Deals, investments, partnerships (part I). PwC Strategy& (2016) S.35	119

Fig. 5.10	The supply side of connected cars: Deals, investments, partnerships (part II). PwC's Strategy& (2016) S.35/36	121
Fig. 5.11	Acquisitions of and investments in autonomous driving capabilities. McKinsey & Company I (2017) S.20	122
Fig. 5.12	Different models of value chain integration. PwC's Strategy& (2017) S.20	124
Fig. 5.13	Roboconomy digital service opportunities. PwC's Strategy& (2017) S.23	125
Fig. 5.14	The full range of connected car technologies and services. PwC's Strategy& (2016) S.15	127
Fig. 5.15	The growth potential of the automotive revenue pool (high-disruption scenario). McKinsey & Company I (2016) S.6	130
Fig. 5.16	Growth of upfront connectivity hardware revenues. McKinsey & Company (2014) S.26	131
Fig. 5.17	Future usage-based revenue potential. McKinsey & Company (2014) S.27	132
Fig. 5.18	Potential of post-warranty revenue re-distribution. McKinsey & Company (2014), S.28	133
Fig. 5.19	Control points and decisive key differentiators. McKinsey & Company (2014), S.33	134
Fig. 5.20	Four ways insurers can use data to improve profitability. A.T. Kearney (2016), S.29	136
Fig. 5.21	Traditional risk-pricing vs. usage-based pricing models. A.T. Kearney (2016) S.35	136

1

Introduction to Automotive Disruption and the Urban Mobility Revolution

1.1 Problem Definition

The automotive industry is standing at the edge of the most incisive transformation since its origin in the late nineteenth century. The degree of private mobility and autonomy that the global diffusion of passenger cars has created, is beyond comparison. Individually accessible means of transport have unlocked an enormous potential, accelerating the pace of societal and economic evolvement. Metaphorically speaking, vehicles have moved humankind closer together. In a vision 2030, this research work addresses the question of how the traditional business model of automotive manufacturers will alter throughout the upcoming decade and how the key drivers of transformation will challenge the present structure of the automotive landscape. Digitalization and the integration of objects into the Internet of Things render possible an increasing dimension of connectivity and data-based vehicle automation levels.[1] Artificial Intelligence (AI) generated by humankind will eventually change the way we move and how we experience mobility.[2] In conclusion, the upheaval

---
[1] cf. A.T. Kearney (2016) How automakers can survive the self-driving era S.33.
[2] cf. BearingPoint (2018) Innovationsradar 2020 S.1.

in the automotive industry will induce a paradigm shift of the archetype of automobile locomotion.

But how will this transformation progress? Which key drivers are crucial to survive this transformation? And how will they impact the traditional business model?

The arms race between the incumbent players has begun. Digitalization and novel forms of data-driven customer experience will emphasize service-centricity as a leading value driver in providing private and shared mobility. In a long-term perspective of autonomous driving, in-motion time could be made usable, thus re-inventing the car as a "third space" between home and work.[3] While there is a common consensus on how the end of the road could look like, nobody can predict the pace and path of the business transformation with absolute certainty.

Therefore, this research work will give a comprehensive insight into the relevance and influence of all stakeholders involved in creating a new mobility ecosystem to provide an exhaustive illustration of the development process.

Beyond an intense cooperation between all stakeholders determining the prospective phenotypes of emerging business models, the success of the automotive revolution is hinged on the question of whether customers are ready to engage in the change. The introduction and commercialization of autonomous driving could be impeded by consumers' inhibition threshold of rendering control to a computer system.[4]

In the end, the human factor will be the pivotal component in the equation. While the technology will eventually reach the required level of maturity, the controversies related to the subject have already spread throughout society.[5] The question is: Who will prevail?

---

[3] cf. Institute for Mobility Research (2018) Autonomous driving—The impact of vehicle automation on mobility behaviour S.24.
[4] cf. The Boston Consulting Group; World Economic Forum (2016) Self-driving vehicles, robot-taxis, and the urban mobility revolution S.8ff.
[5] cf. IBM Center for Applied Insights (2015) Digital disruption and the future of the automotive industry S.4.

1.2 Hyporesearch Work

As indicated in the problem definition, the economic and social significance of vehicle-induced mobility is indisputable. Particularly in Germany, the automotive industry can look retrospectively at a success story steeped in tradition. The direct and indirect proportions of the annual gross value of this economic sector account for 4.5% of the gross domestic product generated in Germany in 2017. Sixty-four percent of the total revenues of 426 billion.[6] Euros were generated in foreign markets making this industry one of the most relevant in terms of global trade and export strength. Beyond the economic importance, the international preeminence of the domestic automotive sector has substantially contributed to the superior global perception of the German industry as a benchmark for high-quality engineered products.[7]

The objective of this research work is to evaluate to what extent the key drivers of the business transformation will impact the existent German automotive industry and how Original Equipment Manufacturers (OEM) can (re-)position themselves to leverage novel business models and to gain access to new revenue sources. In the framework of this evaluation, new phenotypes of mobility scenarios are presented that serve as initial point and guideline for the creation of the business model 2030. In the end, the most decisive question is: Will the German automotive industry be able to defend its preeminence and hold its ground in the competitive environment?

Since automotive disruption is a worldwide phenomenon, this research work intends to take all global influencing factors into consideration that determine the structure of the automobile landscape in 2030. While the survey and analysis have a focal point on the German market, the mobility scenarios and generic assumptions relating to the re-design of the value chain and the redistribution of revenues streams are also transferable to other industrialized nations.[8]

---

[6] Statistisches Bundesamt (2017) S.1.
[7] cf. EY I (2016) Automotive change drivers for the next decade S.2.
[8] cf. PwC - Strategy& (2016) Connected car report 2016 S.35–38.

To provide an in-depth evaluation of the private mobility revolution and the role of automotive manufacturers in the proceedings, the scope of application of autonomous technology in the commercial sector has consciously been taken out of consideration. While advanced mobility solutions bear a vast application potential in the field of logistics and distribution,[9] a thorough assessment of the non-private perspective would have not been possible within the extent of this research work.

1.3 Methodological Structure

This research work intends to present a comprehensive market outlook 2030, based on a qualitative and quantitative research to ensure compliance with the required academic standards. Most of the literature cited within the course of this research work was published by public authorities, business consultancies with specific internal competence centers and automotive associations. The own contribution to the quantitative foundation of the identified key results is provided by the survey conducted to evaluate the prospective usage behavior of autonomous technology functions and the related pace of their market penetration.

To clearly illustrate the impacts of the disruption that will alter the archetype of the automotive industry, the research work commences with a portrayal of the current business models, traditional value chains, and market profile. The subsequent chapter will present the key drivers for automotive disruption and describe the importance of digitalization and data for the viability of advanced mobility concepts and new forms of value creation. To determine if and how the innovations will find their way into the market, survey participants are questioned about their attitude toward connectivity and automated driving, indicating the prospective market volume for autonomous vehicles and related digital services.

Based on a market and customer segmentation, various smart city and mobility scenarios are deducted that provide an insight into the predicted pace and dimension of the market diffusion. After the illustration of the

---

[9] cf. McKinsey & Company I (2017) An integrated perspective on the future of mobility, part 2 S.12.

economic potential induced by automated and shared mobility models, Chap. 5 will focus on the increasing importance of digitalized value chains and the vertical interlinkage with new supplier and distribution networks. The determination of new revenue sources and the identification of distinct use cases will translate the previous qualitative considerations into a financial outlook. The concluding chapter will give a brief wrap-up of the path that lies ahead of the automotive industry and reassess the applicability of the initial assumptions stated in the introduction.

The inclusion of social, economic, financial, and political dimensions will provide the reader with a comprehensive evaluation of the automotive business environment in 2030.

References

BearingPoint. (2018). *Innovationsradar 2020. Welche Trends und Entwicklungen charakterisieren den Handel von Morgen, was sind die Treiber der Innovationen und welche Unternehmen sind dabei besonders innovativ? In: NewRetail (No.1),* S. 1–24. Retrieved March 16, 221, from, https://www.bearingpoint.com/files/NEWretail1_Innovationsradar_2020.pdf&download=1&itemId=496511

EY I. (2016). *Automotive change drivers for the next decade. EY global automotive & transportation sector,* S. 1–16. Retrieved May 1, 2021, from, https://www.ey.com/Publication/vwLUAssets/EY-automotive-change-drivers-for-the-next decade/$FILE/EY-automotive-change-drivers-for-the-next-decade.pdf

IBM Center for Applied Insights. (2015). *Digital disruption and the future of the automotive industry,* S. 1–16. Retrieved June 2, 2021, from, https://www-935.ibm.com/services/multimedia/IBMCAI-Digital-disruption-in-automotive.pdf

Institute for Mobility Research. (2018). *Autonomous driving—The impact of vehicle automation on mobility behaviour,* S. 1–95. Retrieved June 26, 2021, from, https://www.researchgate.net/profile/Stefan_Trommer/publication/312374304_Autonomous_Driving_-_The_Impact_of_Vehicle_Automation_on_Mobility_Behaviour/links/5881fe354585150dde4014a0/Autonomous-Driving-The-Impact-of-Vehicle-Automation-on-Mobility-Behaviour.pdf?origin=publication_detail

Kearney, A. T. (2016). *How automakers can survive the self-driving era. A.T. Kearney study reveals new insights on who will take the pole position in the $560 billion autonomous driving race*, S. 1–36. Retrieved March 16, 2021, from, https://www.kearney.com/automotive/article?/a/how-automakers-can-survive-the-self-driving-era

McKinsey & Company I. (2017). *An integrated perspective on the future of mobility, part 2: Transforming urban delivery.*

PWC – Strategy&. (2016). *Connected car report 2016. Opportunities, risk, and turmoil on the road to autonomous vehicles*, S. 5–63. Retrieved April 27, 2021, from, https://www.strategyand.pwc.com/reports/connected-car-2016-study

Statistisches Bundesamt. (2017). *Wirtschaftskennzahlen.*

The Boston Consulting Group; World Economic Forum. (2016). *Self-driving vehicles, robot-taxis, and the urban mobility revolution*, S. 3–26. Retrieved March 29, 2021, from, https://www.bcg.com/…/automotive-public-sector-self-driving-vehicles-robo-taxis-urban-mobility-revolution.aspx

2

Characterization of the Automotive Industry

2.1 Traditional Business Models of OEMs and Market Profile

In the course of this research work, it will become apparent that the traditional business model of original equipment manufacturers is incrementally modified and extended toward novel forms of mobility provision, altering the archetype of the automotive industry.[1] The business model concept as referred to in countless articles of the economics literature is typically lacking consensus on its definition, solely providing abstract and generic interpretations of the term.[2]

A basic definition describes the business model as a "representation of the value of logic of an organization in terms of how it creates and captures customer value."[3] Further compositional elements that add to the specificity and applicability include the concept of value proposition, the organizational architecture, and financial dimensions. George and

---

[1] cf. PwC; NTT Data Deutschland (2013) Automotive Retail S.11.
[2] cf. Siegfried, Patrick (2017) Strategische Unternehmensplanung in jungen KMU – Probleme und Lösungsansätze S.17–25.
[3] Fielt, Dr. Erwin (2013), Conceptualising Business Models S.85.

Bock refer to business models as the "design of organizational structures to enact a commercial opportunity."[4] In this context, the structure is sub-classified in resource, transactive and value components. The definition of Zott and Amit incorporates further considerations of the value-added process by describing a business model as "a template of how a firm conducts business, how it delivers value to stakeholders […] and how it links factor and product markets."[5] Mass market OEMs' value creation is closely linked with their upstream tier-n supplier network that provides most of the vehicles' components and modules. The value chain in the automotive industry is linear and unidirectional, that is to say, OEMs are standing at the end of the value-added process,[6] distributing the final goods via a sales network of in-house and franchised dealerships to the customers. Hence, automotive producers primarily engage in powertrain production, body manufacturing and final assembly as well as in the provision of spare parts, wholesale and retail finance on the aftersales market.

Before providing a deeper insight into how the automotive landscape is going to change in the perspective 2030, the subsequent sections illustrate the market and competition profile as well as the main challenges, OEMs are currently facing. The set-up of plants, machinery, and equipment in multiple global manufacturing facilities represents vast initial and continuous investment costs leading to a high degree of sunk costs.[7] The complexity of the model and derivative portfolio offered in differing target markets increase the number and variety of production facilities and lines. The growing quantity of sourcing and sales areas simultaneously requires a step-up in warehousing and distribution centers that bundle and allocate component parts provided by an extensive local and cross-border supplier network.[8] The diversification of the product portfolio and the decentralization of production are compensated for by the focus on mass standardization and modularization. Many automotive

---
[4] Bock and George (2017) S.99.
[5] Zott, Christoph; Amit, Raphael (2009) The Network Challenge S.32.
[6] cf. Brucker-Kley, Elke; Kykalova, Denisa; Keller, Thomas (2018) Kundennutzen durch Digitale Transformation S.142.
[7] cf. Oliver Wyman IV (2016) Sourcing in the automotive industry S.3.
[8] cf. Siegfried, Patrick (2020) Handel 4.0 – Erfolgreiche Unternehmenskonzepte mit Arbeitsfragen und Lösungen S.35ff.

manufacturers have developed modular construction systems that aid in achieving economies of scale by decreasing the total number of components used in the assembly of different car models.[9] This procedure allows for the cost-efficiency of the purchasing function and partially decreases the complexity of the supplier structures. Since OEMs need a high utilization of their production capacity to account for the initial and running costs, stable production volumes are a key criterion for the profitability of single manufacturing facilities.[10] The focus on the maximization of sales, however, causes a high inflexibility in the adaption to changes in demand. Therefore, automotive producers are typically granting large budgets to their marketing functions for brand promotion and positioning. The proliferation of OEMs' vehicle portfolio through globalized sales channels has led to high sales incentives that further intensify the fierce price competition[11] and rivalry for market shares.

In view of the altering automobile landscape induced by digitalization and technological evolvement, the figure above illustrates the most decisive trends and market changes in a short-term and long-term perspective. The market saturation of privately owned vehicles in Western and industrialized nations is leading to stagnating sales figures, in turn, aggravating the economic viability of production facilities with excess capacities (Fig. 2.1).[12]

The governmental push for ecological sustainability and the advance of ethical consumerism[13] as key drivers of change are increasingly shifting the business activities toward novel mobility concepts and demotorization. The cost- and time-intensive development of alternative powertrains and the set-up of new mobility services lead to a further deterioration of profit margins. The conversion of the traditional automotive value chain causes a fracturing of existing production plans, supplier networks, and partnerships.[14] With the increasing importance of technological and software-related components, IT companies and digital-native start-ups

---

[9] cf. Natalia, Nefedova (2013) Trends in Automotive Industry S.12–15.
[10] cf. Henneric, Oliver; Licht, Georg (2005) Europe's Automotive Industry on the Move S.24.
[11] cf. Diehlmann, Jens; Häcker, Dr. Joachim (2013) Automotive Management S.175.
[12] cf. Henneric, Oliver; Licht, Georg (2005) Europe's Automotive Industry on the Move S.38.
[13] cf. Diez, Willi (2018) Wohin steuert die deutsche Automobilindustrie? S.14.
[14] cf. Oliver Wyman IV (2016) Sourcing in the automotive industry S.6.

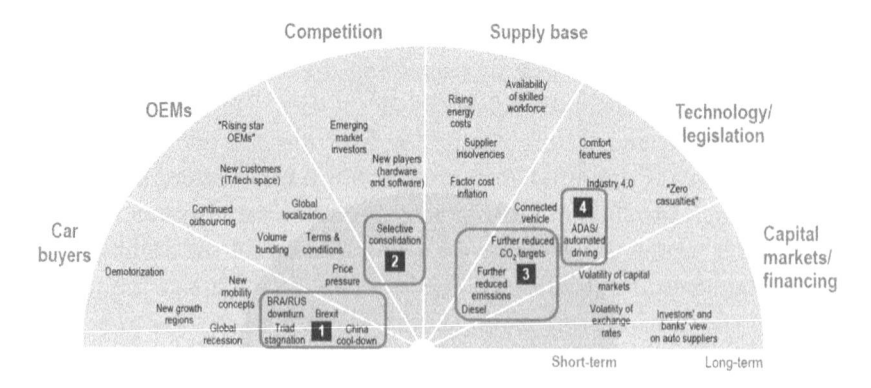

Fig. 2.1 Short-term and long-term trends in the automotive industry. Roland Berger and Lazard (2016), S.17

are claiming a growing share of the total value creation, resulting in a continuous trend of outsourcing and off-shoring toward the new players in the automotive industry.[15] The intensification of the global competition has already introduced a progressive consolidation movement in the supplier industry that need to bundle their resources and core competencies in the fields of technology and software development by merging related company subsidiaries.[16] In particular, small-scale suppliers are threatened by the financial viability and innovation capacity of international IT providers. Beyond the emergence of new, powerful competitors, rising energy costs, factor cost inflation, and the volatility of exchange rates contribute to the rising number of supplier insolvencies. From a long-term perspective, OEMs that fail to adjust to the new market conditions and are not capable of keeping pace with leading providers of connected vehicles, ADAS features, and autonomous technology, will fall victim to the uprising automotive incumbents and new market entrants.[17]

---

[15] cf. Siegfried, Patrick (2013a) The importance of the service sector for the industry S.13–23.

[16] cf. McKinsey & Company IV (2015) Vernetzte Autos und autonomes Fahren als Wachstumstreiber für Zulieferer S.2–4.

[17] cf. McKinsey & Company IV (2016) Performance and Disruption S.9.

2.2 Gearing Up the OEM–Supplier Interface

With the increasing shift from product-centric to service-centric mobility,[18] automotive manufacturers are forced to integrate alternative powertrains, interconnected applications, and advanced technological modules into their supply portfolio. The complexity of supplier networks is thus intensified by the demand for software-based components beyond the mere acquisition of hardware parts.[19]

Before illustrating the prospective importance of digital value chains and corresponding cooperations with IT companies and digital-native start-ups, the subsequent sections depict the traditional OEM–supplier interface prior to the automotive upheaval and presents levers to strengthen the collaborative network for future requirements (Fig. 2.2).

The supplying industry has faced a comprehensive wave of mergers and acquisitions throughout the past decades leading to a consolidation of the upstream business markets.[20] Recent mergers include the coalescence of the Magna International Group with Steyr-Daimler-Puch and the Donnelly Corporation or ZF Friedrichshafen's acquisition of LTD Parts.[21] This consolidation reduces the variety of tier-n suppliers, forcing OEMs to establish strong relationships with its main partners, especially in leading-edge areas of mechatronics and electronics. As suppliers are increasingly becoming innovation drivers, automotive manufacturers need to incrementally participate in R&D cooperations to provide a sufficient level of product differentiation while avoiding over-dependence by proactively engaging in scouting and developing new value-added sources.[22]

The suppliers' share of total value creation is continuously growing due to their emerging role as system integrators. The increasing modularization of vehicles gives the opportunity to provide completely assembled and tested systems posing a challenge for the purchasing function of

---

[18] cf. Kreutzer, Ralf; Neugebauer, Tim; Pattloch, Annette (2017) Digital Business Leadership S.127.
[19] cf. Heneric, Oliver; Licht, Georg (2005) Europe's Automotive Industry on the Move S.29.
[20] cf. Nieuwenhuis, Paul; Wells, Peter (2015) The Global Automotive Industry S.14.
[21] cf. World Economic Forum (2016) Digital Transformation of Industries S.25.
[22] cf. PwC - Strategy& (2017) The 2017 Strategy& Digital Auto Report S.34.

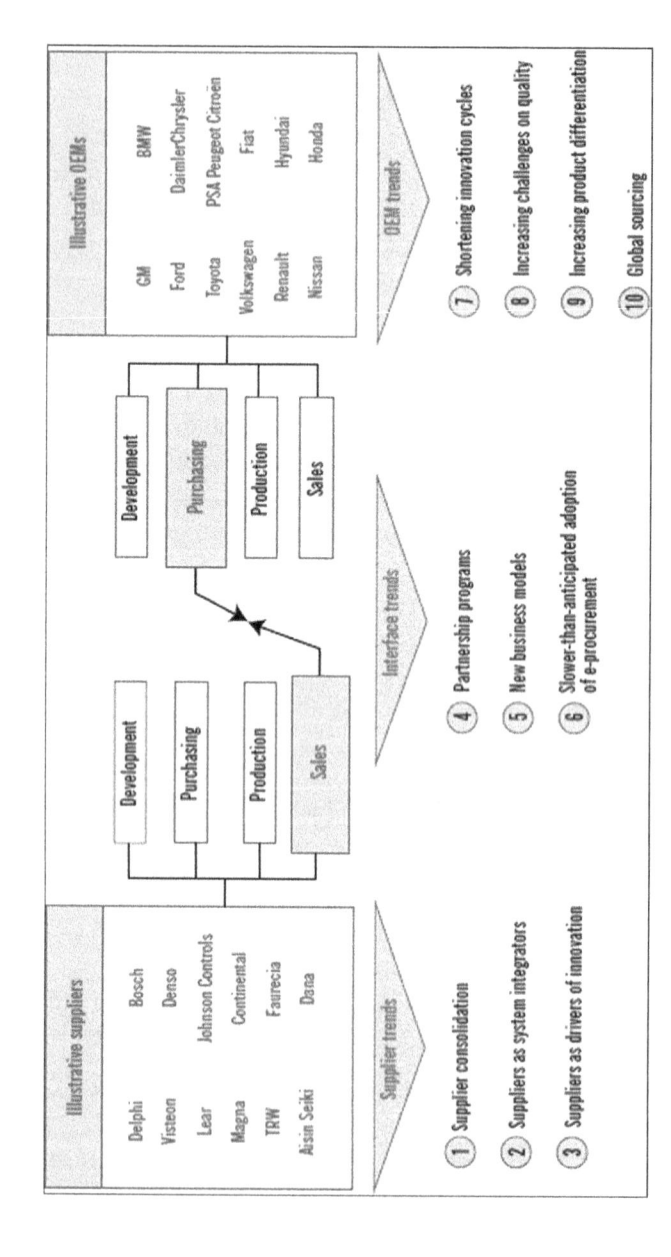

Fig. 2.2 Key trends affecting the OEM–supplier relationship. The Boston Consulting Group (2004), S.12

both, OEMs and suppliers.[23] An accurate evaluation of multipartite modules requires technical and engineering know-how that qualifies to price those complex component systems. As the suppliers' share of added value is increasing, the network of tier-2 and tier-3 suppliers is getting more complex. To ensure a target-oriented identification and integration of upstream partners into their production plans and logistics streams, tier-1 suppliers need to orchestrate a growingly extensive supplier portfolio.[24]

The trend of applying modular design principles simultaneously increases the share of outsourced, external value creation in the automotive industry. Since considerable portions of the vehicle production are shifted to the supplier level, upstream hardware and software developers are taking a leading role in the provision of innovation drivers. Advanced electronics and software applications, in particular, are decisive in the commercialization of new business models. Recent inquiries have shown that the number of patents registered by suppliers is already exceeding those of OEMs significantly, reflecting their strategic importance in developing advanced mobility. To capture profitable business relationships, automotive manufacturers need to provide coordinated innovation processes as well as physical and virtual innovation platforms to facilitate a target-oriented integration of suppliers' market- and product-related expertise.[25] Supplier, in turn, need to focus on leveraging the full potential of their upstream value chains by forming comprehensive innovation networks with tier-n partners.[26]

To achieve long-term collaboration with its supplier base, OEMs have introduced partnership programs that aim at raising quality, strengthening innovative capacities, and cutting shared costs. Those networks could provide an effective platform for innovation exchange because participating suppliers are conceded a genuine privileged status[27] that includes preferential treatments in negotiations and development opportunities.

---

[23] cf. Lazard, Roland Berger (2016): Global Automotive Supplier Study 2016 S.7 10.
[24] cf. The Boston Consulting Group (2004) Beyond Cost Reduction S.7.
[25] cf. The Boston Consulting Group (2004) Beyond Cost Reduction S.10.
[26] cf. Hood, Ray; Kuglin, Fred (2009) Using Technology to Transform the Value Chain.
[27] cf. Wormald, John; Maxton, Graeme (2004) Time for a Model Change S.138.

Cooperative agreements in R&D, production-cycle supply guarantees, or volume-sensitive pricing could further enhance business stability and reduce associated risks.[28]

New phenotypes that are based on the collocation between OEMs and suppliers include Build-Operate-Transfer (BOT) respectively pay-on-production models.[29] In this method, suppliers inherit parts of the production process providing the opportunity to gain deeper expertise, enhance the product development and, thus, realize potentially higher margins. However, BOT models clearly shift operational and market risks toward the supplier possibly compromising their financial viability. Automotive manufacturers benefit from an omission of initial investments that constrain budgets and impact the balance sheet. With the increasing outsourcing and modularization in mind, BOT concepts could result in a further loss of operations competence and increased prices since suppliers typically do not have access to the same low-interest financing conditions. Another model introduced by OEMs to optimize Just-in-Time (JIT) and Just-in-Sequence (JIS) production processes is supplier parks.[30] These simplify the logistical connections with upstream suppliers based on geographical proximity but impede the creation of economies of scale and require unilateral investment from the supplier side.

To reduce complexity in purchasing, OEMs have installed public and private marketplaces such as Covisint or SupplyOn with the intent to create e-procurement platforms,[31] in particular for commodities that do not require extensive interaction between OEMs and suppliers in terms of product specifications or engineering requirements.[32] But the adoption of these platforms is still in an early stage since many modules and components are not sufficiently suited for online bidding and the synchronization and compatibility of IT systems is a cumbersome process. Especially with respect to future commodities in the field of electronics

---

[28] cf. Kallina, Dennis; Siegfried, Patrick (2021) Optimization of Supply Chain Network, S.45–48.
[29] cf. The Boston Consulting Group (2004) Beyond Cost Reduction S.8–12.
[30] cf. Maxwell, Gordan; Drummond, Stuart (2010) Automotive Industry S.112.
[31] cf. Meier, Andreas; Stormer Hendrik (2012) eBusiness & eCommerce S.74.
[32] cf. The Boston Consulting Group (2004) Beyond Cost Reduction S.9–10.

and software-based features, the platforms could be applied as a cooperative development[33] tool, facilitating the exchange of information throughout the engineering process under mutual reconciliation of product requirements.

The increased collaboration of OEMs and supplier networks via BOT models, supplier parks, or e-procurement platforms is primarily owed to the continuous shortening of innovation and product cycles that is yet to be fortified in view of the shift from product-centric to service-centric mobility since relevant software components are subject to even shorter upgrade cycles. In consequence, automotive manufacturers and suppliers need to jointly orchestrate their design and production plans to determine clear interfaces in their overlapping value-added chains.

Pressure on innovation capacity and persistent cost pressure passed on from OEMs to upstream suppliers have led to an increased number of quality defaults and recall actions in the past. However, the licensing and success of autonomous vehicles will be highly dependent on rigorous safety regulations and testing procedures[34] imposed by governmental entities. Therefore, OEMs have already started proactive initiatives dispatching quality- and process-improvement teams to supplier facilities to ensure a consistent quality level.

The increasing product differentiation triggered by customers' demand for vehicles adapted to various usage types and lifestyle requirements, poses an additional challenge to the cost competitiveness and modularization of vehicles. New models and derivatives with unstandardized components impede economies of scale in designing and producing those parts and increase the complexity of the procurement function.[35] In view of high investments in e-mobility and automation, OEMs need to further develop common-part strategies to bundle supply streams in cross model agreements, thus creating financial capacities for R&D budgets.

With the coverage of new sales market for prospective mobility solutions and digital services,[36] the significant value of global sourcing is con-

---

[33] cf. MacGregor, Steven; Torres-Coronas, Teresa (2007) Higher Creativity for Virtual Teams S.279.
[34] cf. KPMG I (2018) Autonomous Vehicle Readiness Index S.22.
[35] cf. Nitschke, Christian (2005) Outsourcing vs. Insourcing in the Automotive Industry S.16.
[36] cf. Schaeffer, Eric (2017) Industry X.0. Digitale Chancen in der Industrie nutzen.

tinuously increasing beyond a mere option for low-cost procurement. Most mass market automotive manufacturers have already entered strategic alliances or established joint ventures with Chinese partners to receive access to the protectionist market.[37] The fragmentation of commercialization of highly automated and autonomous vehicles and related digital features due to differing regulations and infrastructure development stages[38] forces OEMs to closely cooperate with local automotive players and stakeholders to create market shares and quick penetration.

With the emergence of new business models and the upheaval of the automotive industry, the developments described above will further intensify with regard to the diffusion of advanced technologies or the introduction of novel mobility models as illustrated in Chaps. 3 and 4. To prepare for these challenges, the subsequent section presents various levers to optimize the OEM–supplier interface that can also be transferred to collocations with IT companies and start-ups gaining importance in the creation of digital-based business models (Fig. 2.3).

In general, OEMs can pursue two approaches within the framework of procurement that impact the level of product differentiation and brand positioning. The volume approach includes a large network of price-competitive suppliers providing mass standardized components at low cost. However, the short-term perspective underlying this method, impedes an exchange of innovation and leading-edge technology thus limiting opportunities for brand differentiation. In contrast, the creation of innovation-purchasing processes is typically focused on a small number of suppliers that are integrated into the series development, ramp-up, and series production.[39] The long-term oriented, close interlinkage of value-added chains supports the generation of supplier loyalty and innovation drive. Since brand positioning and product differentiation are becoming a crucial value driver for customers, the degree of stabilization of the innovation collaboration and exchange represents a prerequisite in launching state-of-the-art vehicles.

---

[37] cf. PwC - Strategy& (2016) Connected car report 2016 S.34–35.
[38] cf. Anderson, James et al. (2016) Autonomous Vehicle Technology S.138.
[39] cf. Pollak, Dale (2017) Like I See It S.82–84.

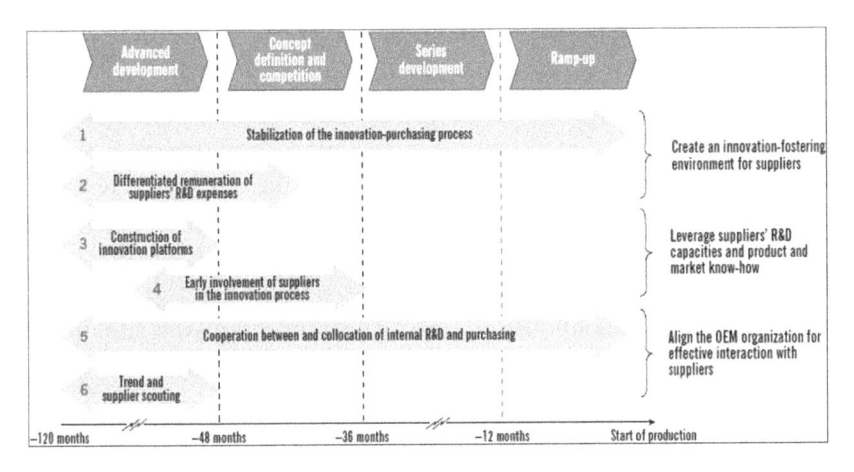

Fig. 2.3 Six levers to optimize the OEM–supplier interface. The Boston Consulting Group (2004), S.20

The creation of innovative capacity can further be improved by remunerating suppliers' R&D expenditures. Basic R&D contracts[40] grant the supplier an adequate compensation for the conduction of specific development tasks without a predetermined link to the subsequent series production.

The next level of collaboration is R&D partnerships in which both parties share personnel and equipment for particular development functions, creating a basis for an efficient transfer of expertise and lowering unilateral investments. The supplier is accorded a preferential status when choosing partners for the launch of the serial production. With regard to innovations as the most significant value and price drivers, production-cycle supply contracts bear the most promising potential for future collaboration forms.[41] Single sourcing of one supplier throughout the entire cycle guarantees a high level of exclusivity of the OEM–supplier business relationship,[42] thus facilitating the access and exchange of generated

---

[40] cf. The Boston Consulting Group (2004) Beyond Cost Reduction S.21.
[41] cf. Bain & Company (1999) The dawn of the mega-supplier S.4–6.
[42] cf. The Boston Consulting Group (2004) Beyond Cost Reduction S.16.

innovations. This approach is primarily suitable for developments in the field of Information Technology (IT) and electronics (Fig. 2.4).[43]

With the emergence of start-ups and small digital players, the establishment of innovations platforms is regaining its momentum.[44] These open web portals are institutionalized opportunities for low-scale companies[45] to submit their innovations and concepts to OEMs' research teams. This virtual infrastructure has also defined regulator frameworks regarding intellectual property, supplier compensation, and potential co-development agreements. This channel represents a win-win situation for both, OEMs and small suppliers in terms of access to innovative concepts and financial remuneration.

Especially in the field of non-traditional innovations, an early involvement of suppliers is key to install a successful joint-development process.[46] Participation in the concept definition or in prior initiation phases facilitates the development of key differentiating features based on a deep

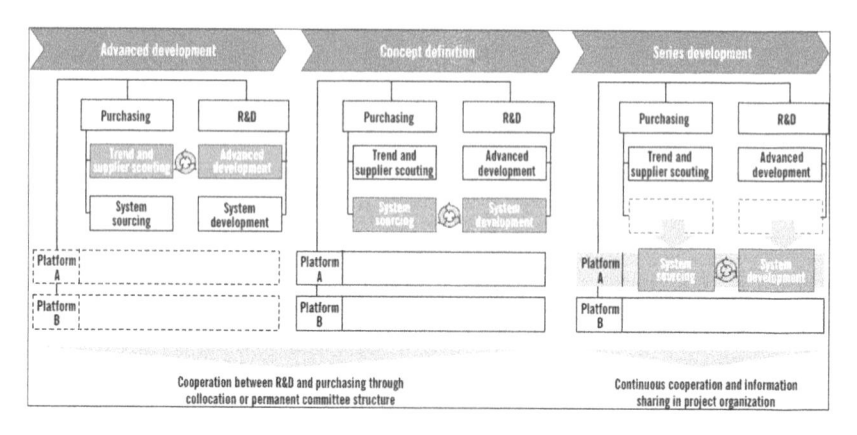

Fig. 2.4 Collocations contribute to an effective cooperation between purchasing and R&D. The Boston Consulting Group (2004), S.25

---

[43] cf.Siegfried, Patrick (2013b) The service engineering concept for business S.173–187.
[44] cf. Siegfried, Patrick (2021a), Business Management Case Studies S.10–15.
[45] cf. Attias, Danielle (2017) The Automobile Revolution S.21.
[46] cf. The Boston Consulting Group II (2016) The Factory of the Future S.5.

understanding of customers' needs and market expertise on the supplier side.

As indicated in the illustration on the previous page, the collocation between the internal R&D and purchasing functions needs to be ensured throughout the entire development cycle.[47] The dilemma between a push for cost reduction on the one hand and sufficient budget sizes on the other hand typically hinder a balanced reconcilement between both parties, leading to costly iterations.[48] A superordinate committee that links both functions and ensures an effective and binding decision-making process could contribute to a more frictionless innovation process and a clear communication with external suppliers.

The last lever to facilitate an early identification of innovation drivers and foster their integration into vehicle development is trend and supplier scouting. Beyond innovation platforms and open web portals, procurement entities need to evaluate potential suppliers in a medium- and long-term vision.[49] The number and pace of novel technologies diffusing into the automotive market requires a distinct selection of the most promising innovation in early development stages. The IT tools and technological advances presented in the next chapter depict a non-exhaustive compilation of such innovations.

2.3 Key Trends Affecting the Mobility Market

Beyond advancements in technological capacities, the introduction and commercialization of alternative mobility solutions have gained momentum by evolving population demography, ethical consumerism, and a diversification of product and market core areas.

The demographic change in Germany is incrementally causing a reduction of the most important customer segments as measured by the average purchasing power.[50] The baby-boomer generation representing the

---

[47] cf. Lange, Christian (2013) Kooperationen in Forschung Entwicklung.
[48] cf. Oliver Wyman IV (2016) Sourcing in the automotive industry S.6.
[49] cf. Bain & Company (1999) The dawn of the mega-supplier S.4.
[50] cf. Roland Berger I (2016a) Automotive 4.0 - Outlook and Implications S.11.

most profitable consumer group in terms of quantity of sales[51] and average profit margin per purchase will reach retirement age in the mid-term. The continuous urbanization and value shifts of younger target groups are additionally affecting the profit structures and rate of returns of mass market producers. As a result of urban growth, the demand for small, cost-efficient, and thus low-profit vehicles is increasing.[52] Public transport and emerging mobility ecosystems are serving the demand for personal locomotion, diminishing the necessity of privately owned vehicles.[53] Especially in saturated markets, ecological sustainability and demotorization trends have replaced traditional values associated with private vehicle ownership.[54]

From a long-term perspective, the automotive core business will further shift toward growth markets, in particular in Asia and South America.[55] The relocation of production facilities and distribution channels will, in consequence, put domestic employment at risk. The increasing share of electric and plug-in vehicles[56] in the product portfolio of automotive manufacturers could further lead to a dislocation of the current value chains, reducing the significance of the German industrial location. The subsequent chapters will further illuminate the indicated changes and argue their impact on the domestic automotive industry and related business models in detail.

References

Anderson, J., Kalra, N., Stanley, K., Sorenson, P., Samaras, C., & Oluwatola, O. (2016). *Autonomous vehicle technology. A guide for policymakers*. Rand Corporation.

---

[51] cf. Iskander Business Partner (2016) Digitalisierung in der Automobilindustrie S.12.
[52] cf. KPMG II (2017) Global Automotive Executive Survey 2017 S.42–44.
[53] cf. Roland Berger (2011) Automotive landscape 2025 S.20.
[54] cf. McKinsey & Company IV (2016) Performance and Disruption S.20.
[55] cf. Siegfried, Patrick (2021b) Land & Sea Transport – Aviation Management S.32–45.
[56] cf. Hülsmann, Michael; Fornahl, Dirk (2014) Evolutionary Paths towards the Mobility Patterns of the Future S.39–40.

Attias, D. (2017). *The automobile revolution. Towards a new electro-mobility paradigm*. Springer Int. Publishing AG.

Bain & Company. (1999). *The dawn of the mega-supplier. Winning supplier strategies in an evolving auto industry*, S. 1–8. Retrieved March 9, 2021, from, http://www.bain.com/publications/articles/Copy_of_Perform_improv_templates.aspx

Berger, R., & Lazard. (2016). *Global Automotive Supplier Study. Being prepared for uncertainty*, S.3-41. Retrieved May 28, 2021, https://www.rolandberger.com/publications/publication_pdf/roland_berger_global_automotive_supplier_2016_final.pdf

Bock, A. J., & George, G. (2017). *The business model book. Design, build and adapt business ideas that driver business growth* (3. Aufl.) Pearson Education Limited.

Brucker-Kley, E., Kykalova, D., & Keller, T. (2018). *Kundennutzen durch Digitale Transformation. Business-Process-Management-Studie—Status Quo und Erfolgsmuster*. Springer Gabler.

Diehlmann, J, & Häcker, J. Dr. (2013). *Automotive management. Navigating the next decade* (2. Aufl.) München: Oldenbourg Verlag.

Diez, W. (2018). *Wohin steuert die deutsche Automobilindustrie?* (2. Aufl.) Walter de Gruyter GmbH.

Fielt, E. Dr. (2013). Conceptualising business models: Definitions, frameworks and classifications, S.85-105, *Journal of Business Models,* 1(1). Retrieved May 17, 2021, from, https://journals.aau.dk/index.php/JOBM/article/download/706/543

Henerie, O., & Licht, G. (2005). *Europe's automotive industry on the move. Competitiveness in a changing world* (p. 32). Physica Verlag.

Hood, R., & Kuglin, F. (2009). *Using technology to transform the value chain*. Taylor & Francis Group.

Hülsmann, M., & Fornahl, D. (2014). *Evolutionary paths towards the mobility patterns of the future*. Springer Verlag.

Iskander Business Partner. (2016). *Digitalisierung in der Automobilindustrie. Wer gewinnt das Rennen? Traditioneller Automobilhersteller oder Silicon Valley?* S. 3–28. Retrieved July 3, 2021, from, http://i-b-partner.com/wp-content/uploads/2016/08/2016-09-06-Iskander-RZ-Whitepaper-Digitalisierung-in-der-Automobilindustrie-DIGITAL.pdf

Kallina, D., & Siegfried, P. (2021). Optimization of supply chain network using genetic algorithms based on bill of materials. *The International Journal of Engineering & Science*. https://doi.org/10.9790/1813-1007013747

KPMG I. (2018). *Autonomous vehicle readiness index. Assessing countries' openness and preparedness for autonomous vehicles*, S. 1–60. Retrieved July 22, 2021, from, https://assets.kpmg.com/content/dam/kpmg/nl/pdf/2018/sector/automotive/autonomous-vehicles-readiness-index.pdf

KPMG II. (2017). *Global automotive executive survey 2017*, S. 1–56. Retrieved May 6, 2021, from, https://assets.kpmg.com/content/dam/kpmg/es/pdf/2017/global-automotive-executive-survey-2017.pdf.

Kreutzer, R., Neugebauer, T., & Pattloch, A. (2017). *Digital business leadership. Digital transformation, business model innovation, agile organization, change management.* Springer Fachmedien GmbH.

Lange, C. (2013). *Kooperationen in Forschung Entwicklung. Die Automobilindustrie - Vorbild für andere Branchen?* Diplomica Verlag GmbH.

MacGregor, S., & Torres-Coronas, T. (2007). *Higher creativity for virtual teams. Developing platforms for co-creating.* IGI Global.

Maxwell, G., & Drummond, S. (2010). *Automotive industry. Technical challenges, design issues and global economic crisis.* Nova Science Publishers.

McKinsey & Company IV. (2015). *Vernetzte Autos und autonomes Fahren als Wachstumstreiber für Zulieferer.* Hattrup, Martin. Retrieved March 22, 2021, from, https://www.mckinsey.de/vernetzte-autos-und-autonomes-fahren-als-wachstumstreiber-fuer-zulieferer

McKinsey & Company IV. (2016). *Performance and disruption. A perspective on the automotive supplier landscape and major technology trends*, S. 1–32. Retrieved March 22, 2021, from, http://worldmobilityleadershipforum.com/wp-content/uploads/2016/06/160324-McKinsey-Supplier-Brochure-Performance-and-Disruption.pdf

Meier, A., & Stormer, H. (2012). *eBusiness & eCommerce. Management der digitalen Wertschöpfungskette* (3. Aufl.) Springer Verlag.

Natalia, N. (2013). *Trends in automotive industry: New mobility concept. Rethinking current business models of OEMs 2013*, S. 1–96. Retrieved June 2, 2021, from, http://www.makingsciencenews.com/catalogue/papers/217/download

Nieuwenhuis, P., & Wells, P. (2015). *The global automotive industry* (1. Aufl.). John Wiley & Sons, Ltd.

Nitschke, C. (2005). *Outsourcing vs. insourcing in the automotive industry—The role and concepts of suppliers. The rise of mega suppliers* (1. Aufl.) GRIN Verlag, Open Publishing GmbH.

Oliver Wyman IV. (2016). Sourcing in the automotive industry: How can suppliers create more value? In *Insights on automotive supplier excellence*, S. 2–10.

Retrieved April 23, 2021, from, http://www.oliverwyman.com/content/dam/oliver-wyman/v2/publications/2016/jan/OliverWyman_Sourcing_in_the_AutomotiveIndustry_web.pdf

Pollak, D. (2017). *Like I see it. Obstacles and opportunities shaping the future of retail automotive.* vAuto Press.

PWC, NTT Data Deutschland. (2013*). Automotive retail—Die Zukunft beginnt jetzt,* S. 12–56. Retrieved June 6, 2021, from, https://www.pwc-wissen.de/pwc/de/shop/publikationen/Automotive+Retail+-+Die+Zukunft+beginnt+jetzt!/?card=13001

PWC - Strategy&. (2016). *Connected car report 2016. Opportunities, risk, and turmoil on the road to autonomous vehicles,* S. 5–63. Retrieved April 27, 2021, from, https://www.strategyand.pwc.com/reports/connected-car-2016-study

PWC - Strategy&. (2017). *The 2017 strategy& digital auto report. Fast and furious: Why making money in the "roboconomy" is getting harder,* S. 1–41. Retrieved April 27, 2021, from, https://www.strategyand.pwc.com/reports/fast-and-furious

Roland, B. (2011). *Automotive landscape 2025: Opportunities and challenges ahead,* S. 1–47. Retrieved July 12, 2021, from, http://www.forum-elektromobilitaet.ch/fileadmin/DATA_Forum/Publikationen/Roland_Berger_2011_Automotive_Landscape_2025_E_20110228.pdf

Roland, B. (2016a). *Automotive 4.0—Outlook and implications. TRAN hearing on "Towards a European Road Safety Area",* S. 1–19. Retrieved April 23, 2021, from, http://www.europarl.europa.eu/cmsdata/99142/9_Presentation%20Sebastian%20Feldmann_Roland%20Berger%20GmbH.pdf

Schaeffer, E. (2017). *Industry X.0. Digitale Chancen in der Industrie nutzen* (1. Aufl.) RedLine Verlag.

Siegfried, P. (2013a). *The importance of the service sector for the industry, Teaching Crossroads: 9th IPB Erasmus Week.* Instituto Politécnico de Braganca.

Siegfried, P. (2013b). The service engineering concept for business, Entrepreneurship-Conference, University of Lisboa/PRT.

Siegfried, P. (2017). Strategische Unternehmensplanung in jungen KMU - Probleme and Lösungsansätze, de Gruyter/Oldenbourg Verlag.

Siegfried, P. (2020). *Handel 4.0—Erfolgreiche Unternehmenskonzepte mit Arbeitsfragen und Lösungen.*

Siegfried, P. (2021a). *Business management case studies.* BoD.

Siegfried, P. (2021b). *Land & sea transport—Aviation management.* BoD.

The Boston Consulting Group. (2004). *Beyond cost reduction. Reinventing the automotive OEM-supplier interface,* S. 3–48. Retrieved April 23, 2021, from, https://www.bcgperspectives.com/content/articles/automotive_sourcing_procurement_beyond_cost_reduction_reinventing_automotive_oem_supplier_interface/

The Boston Consulting Group II. (2016). *The factory of the future*, S. 2–16. Retrieved June 22, 2021, from, https://www.bcg.com/de-de/publications/2016/leaning-manufacturing-operations-factory-of-future.aspx

World Economic Forum. (2016). Digital transformation of industries. Automotive industry. In *Digital enterprise*, S. 1–45. Retrieved May 11, 2021, from, http://reports.weforum.org/digital-transformation/wp-content/blogs.dir/94/mp/files/pages/files/digital-enterprise-narrative-final-january-2016.pdf

Wormald, J., & Maxton, G. (2004). *Time for a model change. Re-engineering the global automotive industry*. Cambridge University Press.

Zott, C., & Amit, R. (2009). *The network challenge. Strategy, profit and risk in an interlinked world*. Wharton School Publishing.

3

Elements of the Automotive Disruption

3.1 Disruptive Mobility and IT Trends

3.1.1 Connectivity and the Internet of Things

As consumers are increasingly getting interconnected and are participating in a globally linked society via mobile smart devices, companies are forced to shift their business activities progressively into a new format of customer interaction. Existing business models are incrementally being challenged by emerging data-based players pushing innovative digital services into the market[1] that are configured to directly appeal to differentiated customer target groups.[2] Privately owned cars are progressively linked to higher Total Costs of Ownership (TCO) and are losing their former association with individual status and personal mobility guarantor. New mobility concepts, that will be depicted in the subsequent chapters, are shifting the perception of the car from a product-centric to an increasingly customer-centric mode of transport.[3] The new challenge,

---

[1] cf. ESCP Europe (2017) Automotive Disruption Switch to Digital Services S.7–10.
[2] cf. Bundesverband Digitale Wirtschaft (2016) Connected Cars—Geschäftsmodelle S.2.
[3] cf. Capgemini Consulting (2017) Beyond the Car S.26–27.

even beyond the automotive industry, is to build client intelligence to expand the quantity of digital touchpoints on the customer journey. The way companies address and interact with its customers' needs to be personalized to match their own mode of living, product preferences, and mobility requirements.[4] Automotive manufacturers and providers of digital mobility-related services are confronted with the task to exploit the data that accrues during the customer life-cycle and transform it into valuable data for their business activities.[5] The analysis and processing of this data facilitates drawing conclusions that impact the whole value chain from an optimized procurement function to personalized service offerings in the aftersales market. To achieve this data intelligence and develop new business models, the exploitation, targeted analysis, and comprehensive management of vast data volumes, generally related to as big data, is a major requirement.[6] The economic and intelligent usage of the comprised information content enables the optimization of commercial patterns based on big data algorithms. The traditional way of managing complexity through experience and good intuition is increasingly replaced by automated decision processes that are driven by value chain and customer insight.[7] Premise for capturing and gathering any vehicle and user-specific data is the integration of the car into the Internet of Things (IoT).[8]

Technical and environmental sensors, cameras, and microphones are used to monitor the vehicle's performance status, track malfunctions, and collect utility-related data.[9] Processing this data requires onboard data storage serving as a local hardware repository for generated data flows. In order to handle the gathered data in real-time, advanced developments in connectivity and cloud-based features that link the vehicle and its devices with the OEMs data servers and processing software are a significant step

---

[4] cf. Anderson, James et al. (2016) Autonomous Vehicle Technology S.75ff.
[5] cf. Siegfried, Patrick (2021) Enterprise Management Automobile Industry Business Cases S.77–83.
[6] cf. Blanke, Tobias (2014) Digital Asset Ecosystems S.87.
[7] cf. The Boston Consulting Group I (2016) Digital reroutes the auto purchase journey and OEM strategies S.6.
[8] cf. Friess, Peter (2014) Internet of Things.
[9] cf. KPMG I (2016) I See. I Think. I Drive. (I learn) S.10.

to retrieve an additional value.[10] Gateways of this high-speed data connectivity include, amongst others, wireless internet, RFID technologies, Bluetooth, and 5G networks that bridge the geographical gap between the onboard and outbound data storage.[11] The Internet of Things thus not only incorporates vehicles but is a general term for all devices that are integrated into the internet via wireless connections and cloud platforms.[12] This enables a Vehicle-to-Everything (V2X)[13] communication, a link between the vehicle and all stakeholders such as traffic participants, digital service providers, smart infrastructure components, and companies directly and indirectly involved in using the car as a platform for business activities. The interconnectedness of the vehicle and the creation of a digital network[14] is the headstone for the viability of future business models, new mobility concepts and, thus, the transformation of the automotive industry.

3.1.2 A Foresight into the Era of Autonomous Driving

The concept of motorized driving, as people have known it since its invention in the late eighteenth century, is seeing an imminent revolution that is changing human mobility in a disruptive way. The era of autonomous driving has been initiated by the successive development of connectivity features that enable advanced driving assistance features and the embedment of the vehicle into a digital network, establishing possibilities of real-time traffic alerts and navigation, lane departure and automated collision warnings, adaptive cruise control or assisted parking.[15] The emergence and continuous advancements of car features that increasingly relieve the user from his driving function shows a progressive movement towards an era in which drivers are becoming passengers. In this

---

[10] cf. McKinsey & Company III (2016) Monetizing Car Data S.8.
[11] cf. European Commission (2017) Digital Transformation Monitor S.2ff.
[12] cf. Müller, Stefan (2016) Internet of Things (IoT). Ein Wegweiser durch das Internet der Dinge.
[13] cf. Karls, Ingolf; Mueck, Markus (2018) Networking Vehicles to Everything.
[14] cf. Linnhoff-Popien, Claudia; Schneider, Ralf (2018) Digital Marketplace unleashed S.479.
[15] cf. Jurgen, Ronald K. (2013) Autonomous Vehicles for Safer Driving S.20.

relation, the connected car, whose market penetration is proceeding in all vehicle segments, depicts just the first step of the future mobility scenario.

While the analysis of the potential development, market diffusion scenarios, and use cases will be depicted in detail in Chap. 5, this chapter illustrates the preliminary timeline and implementation roadmap of autonomous cars until 2030 (Fig. 3.1).

The development will not be a single leap, but a steady, gradual, and rapid progression through a series of vehicle generations.[16] The speed of marketability depends on a technology roadmap that comprises advancements in big data analytic tools, high-capacity mobile broadband systems, or artificial intelligence.[17] Governmental regulations that foster or impede the introduction of fully automated vehicles as well as customer demand patterns and the acceptance of new mobility concepts will have an impact on the market penetration[18] of autonomous cars and their diffusion across all urban and rural areas. The process of achieving the goal of autonomous mobility thereby challenges many of today's paradigms around cars (Fig. 3.2).

As autonomous driving reaches a high market share it will lead to societal benefits that cover safety, environmental, and cost issues. Accidents

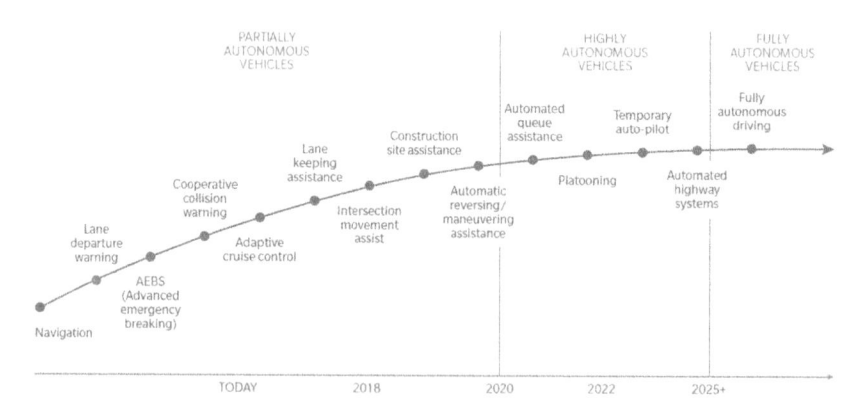

Fig. 3.1 Roadmap towards fully autonomous driving. Wyman, Oliver (2017), S.49

---

[16] cf. Winkelhake, Uwe (2017) Die digitale Transformation der Automobilindustrie S.99–101.
[17] cf. BearingPoint III (2017) Roboter, Rebellen, Relikte S.8.
[18] cf. Deloitte University Press (2017) Governing the Future of Mobility S.9.

3 Elements of the Automotive Disruption

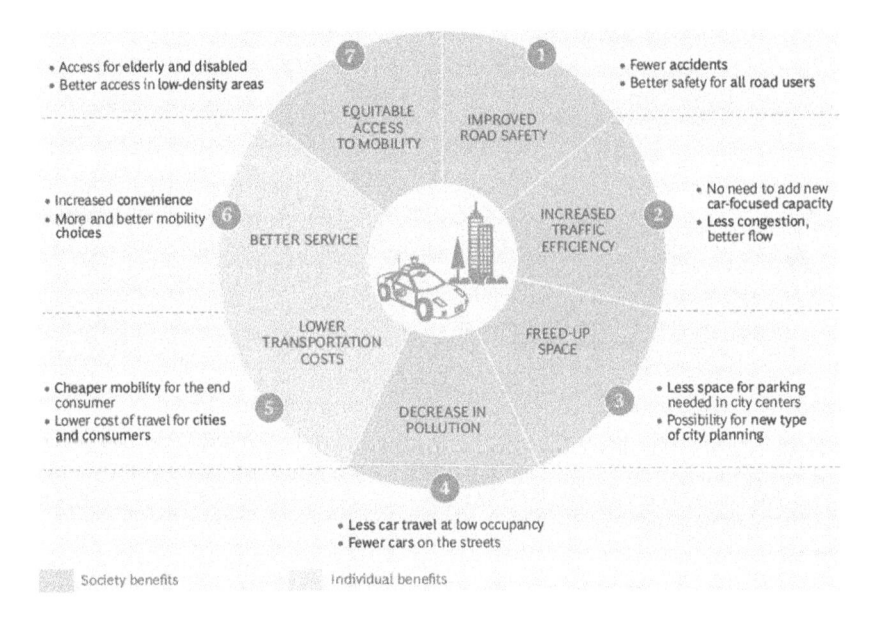

Fig. 3.2 SDV benefits for individuals and society. The Boston Consulting Group/World Economic Forum (2016) S.14

that were formerly caused by human car operators will diminish drastically leading to lower insurance and health care costs.[19] Critical incidents referable to fatigue, drunkenness, insufficient reaction times, or inappropriate safety distances can be eradicated by technically mature smart vehicles.[20] An intelligent infrastructure that communicates with the car as well as a higher capacity utilization and, thus, a decreasing number of vehicles participating in the traffic helps to manage a frictionless traffic flow and optimize the overall energy consumption (Fig. 3.3).[21]

The degree to which autonomous vehicles can contribute to improving the car-related societal burdens is thereby highly dependent on the cooperation of infrastructure planning, OEM investments into Research and

---

[19] cf. KPMG I (2016) I See. I Think. I Drive. (I learn) S.20.
[20] cf. A.T. Kearney (2016) S.4.
[21] cf. Anderson, James et al. (2016) Autonomous Vehicle Technology S.14–16.

Safety	• Real-time **emergency calls** • Early and on-scene **accident information** to support rescue services • **Real-time road hazard warning** (vehicle adaptation towards upcoming road situations)	Time	• **Reduced** customer **delivery time** through optimized routing/navigation and traffic management system • **Reduced time** to find **parking** through networked parking and connected navigation
Convenience	• **Reduced breakdown risk and vehicle downtime** through predictive maintenance (onboard diagnostic) and spare parts management at dealer/workshop • **Increased customer convenience by concierge services** (refueling, carwash, in-trunk delivery) • Improved experience from connected lifestyle	Cost	• **Reduced insurance cost** through PAYD insurance • **Reduced toll/road tax rates** through automated payment infrastructure • **Reduced customer mobility cost** through customer receptiveness to in-car purchasing or exposure to targeted in-car advertising

Fig. 3.3 The increased use of car data will unlock new customer benefits in four areas. McKinsey & Company II, (2016), S.8

Development (R&D)[22] as well as consumer preferences that accelerate the diffusion and implementation of smart mobility.

3.1.3 Urban Mobility Concepts and the Electrification of the Car

As discussed in the previous sections, novel digital tools and the integration of the vehicle into the Internet of Things are paving the way to disrupt the mobility industry. The technology-driven trends can be assigned to four superordinate clusters that comprise electrification, connectivity, autonomous and diverse mobility. The path to the provision of Mobility-as-a-Service (MaaS) models[23] that radically alter the mobility landscape proceeds over five evolutionary levels on which vehicles are iteratively capacitated to conduct highly automated and, eventually, autonomous driving functions. Besides the industry's efforts to digitally interconnect all vehicles, there is a second significant endeavor progressing simultane-

---

[22] cf. Deloitte University Press (2015) Patterns of Disruption S.14.
[23] cf. Wedeniwski; Sebastian (2015) The Mobility Revolution in the Automotive Industry S.267.

3 Elements of the Automotive Disruption

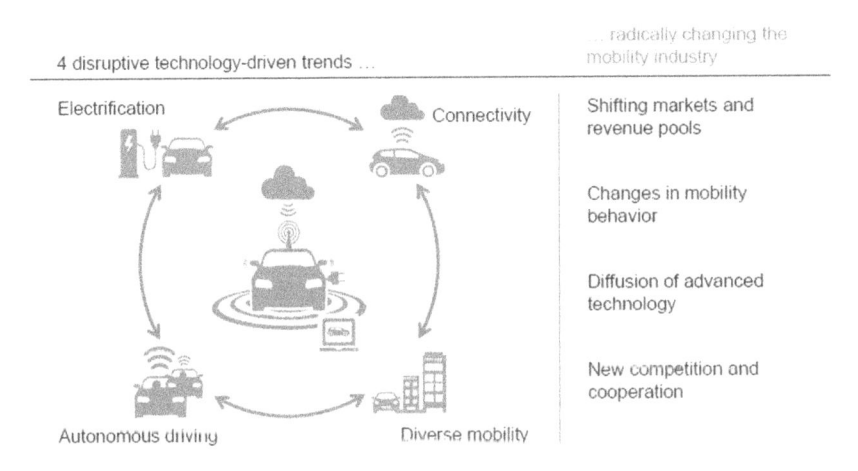

Fig. 3.4 Global trends triggering change in the mobility industry. McKinsey & Company III, (2016), S.10

ously that focuses on the development of alternative power trains, especially hybrids, plug-ins and battery-electrified vehicles (Fig. 3.4).[24]

Electrified Vehicles are not new to the automotive industry. In the early twentieth century, the ancestors of today's EVs already had an appearance but failed to assert their position in the automotive market. For a long time, the competitor of Internal Combustion Engines (ICE) has vanished since its development was not considered profitable due to neglectable customer demand.[25] Higher initial costs and limited battery range coupled with uncertainties regarding infrastructure adjustments and residual values have reinforced the persistent reluctance, degrading Electric Vehicles (EV) to a shelf warmer.[26] However, stricter emission control, decreasing battery costs, extensively available charging stations, and growing customer acceptance create a new momentum for the penetration of electrified vehicles. Fueled by the diesel emission scandal, continued regulator focus on diminishing fine particles and nitrogen oxide (NOx) as well as governmental strategies concerning the reduction of

---
[24] cf. Hulsmann, Michael; Fornahl, Dirk (2014) Evolutionary Paths towards the Mobility Patterns of the Future S.42ff.
[25] cf. UBS (2017) Longer Term Investments. Smart Mobility S.4.
[26] cf. Nieuwenhuis, Paul; Wells, Peter (2015) The Global Automotive Industry S.185.

fossil fuel dependency and global warning, the interest in EVs has spiked.[27] Since the speed of adoption is determined by the equation of consumer pull, mostly driven by the Total Cost of Ownership (TCO),[28] and the regulatory push, adoption rates are likely to be highest in developed, dense cities with strict emission controls and consumer incentives such as discounted electricity pricing, tax abatements or special parking and driving privileges. Beyond that, metropolitan areas show a higher density of charging infrastructure and citizens are less dependent on the possible driving range.

Research suggests that battery cost will potentially decrease to around 150–200 US-Dollars per kilowatt-hour (kWh) by 2025,[29] thus achieving cost competitiveness with conventional cars, creating the most significant catalyst for market penetration. In conclusion, the market success of alternative power trains heavily depends on legislation to drive electric vehicles sales that further triggers the necessity to implement adequate infrastructure and leads to scale effects which, in turn, bring the prices to a level where they can compete with conventional vehicles. The speed of adoption will also vary locally based on the industrial background of the specific countries. Nations with strong automotive industries and a traditional infrastructure typically approach the EV disruption in a less aggressive way, mostly relying on a haphazard network of incentives instead of defining restrictive quotas such as Norway. In contrast, emerging markets, for instance, China, that lack a strong internal combustion engine technology heritage are aggressively attempting to leapfrog other nations by establishing a modern energy infrastructure.[30] For OEMs and suppliers, the ongoing decline of a conventional engine and transmission components depicts a complex challenge. The electronics landscape will most likely experience extensive change as power units and control systems migrate towards a vehicle system electrification. According to a comprehensive study, one-third of the overall value creation of traditional vehicles will substantially change, affecting specific business models and

---

[27] cf. KPMG II (2017) Global Automotive Executive Survey 2017 S.13.
[28] cf. Deutsche Bank AG (2017) The digital car S.25.
[29] cf. Deloitte I (2017) The Future of the Automotive Value Chain S.46.
[30] cf. Hülsmann, Michael; Fornahl, Dirk (2014) Evolutionary Paths towards the Mobility Patterns of the Future S.143ff.

established value chains.[31] Realigning research, development, and procurement functions, re-coordinating the production facilities as well as implementing new marketing and sales channels are posing a burdensome task. Furthermore, EVs could result in obsolescence of established energy delivery infrastructure such as gas stations and trigger new, ideally smart electric-grid-based distribution systems.[32] OEMs as well as their supplier and dealer network need to revise their collaboration models to adapt the traditional value chain to the emerging bestseller.

The actual initial starting point of making autonomous vehicles become a reality is represented by the connected car. A connected car is capable of sharing real-time information with the immediate vicinity by communicating with other vehicles (V2V) or elements of the infrastructure (V2I).[33] This includes the usage of data-based driving functions such as an automated adjustment of vehicle speed to the traffic flow or preventive collision warning.

The next step of the connected car evolution, defined as Level 3/4 of the autonomy scale,[34] indicates a vehicle with automated driving functions that can conduct certain activities without the intervention of a driver.

These functions, among which an autopilot mode, temporary platooning, or autonomous parking on private property can be listed, are enhancing the car ownership and driving experience in terms of convenience, efficiency, and safety. The vehicle does not have to be connected to the cloud since certain automated functions are based on integrated sensory technology and actuating elements. However, these functions do not absolve the driver from his responsibility to be able to control the car in any given situation.[35] The last leap of evolution is a vehicle that drives, steers, accelerates, and bakes entirely autonomously under all driving conditions. This stage of development even capacitates the car to assume driving functions without any human driver onboard such as

---

[31] cf. Bock, Adam J.; George, Gerard (2017) The Business Model Book.
[32] cf. Roland Berger (2017) Automotive Disruption Radar S.12–14.
[33] cf. Winkelhake, Uwe (2017) Die digitale Transformation der Automobilindustrie S.106.
[34] cf. U.S. Department of Commerce, Economics and Statistics Administration (2017) The Employment Impact of Autonomous Vehicles S.4.
[35] cf. Deutsche Bank AG (2017) The digital car S.6.

autonomous trips to the gas station or searching for distant parking spaces.[36] The system uses internal and external information which it acquires through communication with other cars, the infrastructure, or the cloud. The vehicle can complement and update its own sensory data in real-time with information from different sources, e.g., the driving behavior of other cars, traffic supervision or congestion reports. Eventually, since all decisive driving functions are operated by the vehicle systems, the liability regarding vehicle-inflicted accidents is likely to devolve to the car manufacturer or his component suppliers.[37] But this stage is yet to come. "At present, even the smartest cars are pretty dumb. They're engineering marvels, but inflexible marvels." (Fig. 3.5).[38]

As depicted in the table above, connectivity represents a catalyst for the creation of new, digital mobility concepts and business models. Traditional mobility solutions will incrementally give way to new services that provide on-demand and shared mobility in a Peer-To-Peer network (P2P)[39] using mobile applications that integrate the customer into a digital mobility ecosystem. Mobility-as-a-service models have already emerged

	Traditional mobility solutions	New mobility services	
Individual-based mobility	Private car ownership	Car sharing: peer to peer	A peer-to-peer platform where individuals can rent out their private vehicles when they are not in use
	Taxi	E-hailing	Process of ordering a car or taxi via on-demand app. App matches rider with driver and handles payment
	Rental cars	Car sharing: fleet operator	On-demand short-term car rentals with the vehicle owned and managed by a fleet operator
Group-based mobility	Car pooling	Shared e-hailing	Allows riders going in the same direction to share the car, thereby splitting the fare and lowering the cost
	Public transit	On-demand private shuttles	App and technology enabled shuttle service. Cheaper than a taxi but more convenient than public transit
		Private buses	Shared and Wi-Fi-enabled commuter buses available to the public or to employees of select companies. Used to free riders from driving to work

Fig. 3.5 New mobility services offer transportation alternatives. McKinsey & Company II, (2015), S.13

---

[36] cf. EY (2017) The evolution in self-driving vehicles S.6.
[37] cf. Bundesverband Digitale Wirtschaft (2016) Connected Cars—Geschäftsmodelle S.12–14.
[38] cf. PwC - Strategy& (2016) Connected car report 2016 S.55.
[39] cf. Meyer, Gereon; Shaheen, Susan (2017) Disrupting Mobility S.105.

in the form of Uber, DriveNow, car2go, or Lyft[40] that make private car ownership dispensable. The sub-chapters about the creation of the smart city and the scenario analyses will further illuminate current developments and strategies in the mobility markets.

3.2 Digital Transformation

3.2.1 Data—The Emergence of a New Currency

The challenge of maintaining a competitive advantage in this dynamic, innovation-driven business environment is dependent on a new source of value. The trend from product-centric towards service-centric product portfolios keeps shifting the value of assets from tangible forms such as financial funds or capital goods towards intangible assets.[41] Data is emerging as a new strategic asset that is vital to a functional development of a digital value chain, the implementation of smart production as well as the creation of smart products.[42] It is the new leading source of market power in a global digital ecosystem.[43] Data ownership and the ability to collect, process and combine gathered data is a basic prerequisite to deploy customer services and to reengineer the overall value-added process as illustrated in the subsequent chapter. "The combination of autonomy and connectivity will create a third space beyond home and work"[44] to which companies are progressively gaining access. This new personal space could potentially transform the car into a moving digital experience center, a new control point of data generation, and a platform for providing products and services.[45] The car and its user generate a large spectrum of different macro-categories of data that are exploitable for a variety of stakeholders. Infrastructure providers and authorities can utilize data

---

[40] cf. McKinsey & Company II (2015) Urban mobility at a tipping point S.15.
[41] cf. Aldrich, Douglas F. (1999) Mastering the Digital Marketplace S.237.
[42] cf. Siegfried, Patrick (2014) Knowledge Transfer in Service Research—Service Engineering in Startup Companies S.22ff.
[43] cf. Elgar, Edward (2007) The Digital Business Ecosystem.
[44] McKinsey & Company II (2016) Car Data: Paving the Way to Value-creating Mobility S.14.
[45] cf. Aldrich, Douglas F. (1999) Mastering the Digital Marketplace S.27ff.

about external conditions, collected by cameras and sensors to share traffic flow information, feed road usage monitoring systems, and trigger alerts in case of worsening weather conditions.[46] Data about the technical status of the equipment will help OEMs, tier-n-suppliers, and aftersales service providers to improve product specifications, conduct on-board diagnostic processes or inform the user about preventive maintenance options.[47]

Information gathered about the product usage enable the integration of the car with mobility platforms that offer a variety of multi-modal transport possibilities or shared mobility services[48] that create a personalized mobility concept for the user. Personal data and customer preferences are beneficial to retailers and service providers since offerings can be clustered to differentiated customer target groups including content streaming, customized advertising, or concierge services.

The examples provided thereby only show an excerpt of potential use cases and revenue streams that are illustrated in detail in Chap. 5. Data availability and thus the potential to generate value from delivering data-based customer services is highly dependent on building and sustaining customers' trust. Cybersecurity and regulations about data ownership and application rights will become a critical capability for industry players.[49] In this context, the perceived privacy sensitivity varies regarding personal data or equipment-related data.[50] To exploit the full potential of big data analytics and business models dependent on customer insight, the public perception of data security poses a large barrier. Regulators and companies must cooperate and define guidelines and laws to guarantee a secure data ecosystem and the privacy of consumers.[51] This will increase the value of the novel data currency (Fig. 3.6).

---

[46] cf. World Economic Forum (2016) Digital Transformation of Industries S.10.
[47] cf. Linnhoff-Popien, Claudia; Schneider, Ralf (2018) Digital Marketplace unleashed S.297.
[48] cf. Roland Berger (2017) Automotive Disruption Radar S.7–9.
[49] cf. Neckermann, Lukas (2015) The Mobility Revolution S.139–141.
[50] cf. McKinsey & Company II (2016) Car Data: Paving the Way to Value-Creating Mobility S.8–10.
[51] cf. Federal Ministry of Transport and Digital Infrastructure (2015) Strategy for Automated and Connected Driving S.25.

Driver, passengers
(via personal and/or wearable devices)
- Telecommunications (telephone, SMS, e-mail)
- Audio applications/traffic information
- Handheld/portable navigation)

Service providers
- Contents streaming (e.g., audio, video, news, weather)
- Direct mobile payments
- Pay-as-you-drive (PAYD) insurance
- Reservations/concierge services

Mobility providers
- E-hailing services (for cars, LCVs)
- Vehicle sharing
- Public transport hubs (for integrated mobility)

Authorities
- Emergency and breakdown calls
- Law enforcement (for police)
- Vehicle-data-based road maintenance

OEM (and dealers)
- Remote onboard diagnostic and preventive maintenance
- Enhanced product design through "field-data" recovery (actual user data)
- Accurate warranty management system

Infrastructure
- Automated road toll/taxation system
- Average speed monitoring systems
- Traffic flow management and monitoring systems

Other cars
- Rolling map network
- Safety systems (i.e., pre-collision warning thanks to data from other cars)
- Automatic cruise control (incl. lane/distance keeping)

Retailers
- In-car offerings and targeted advertising
- Proximity/customers flow data analytics for store location, opening hours optimization

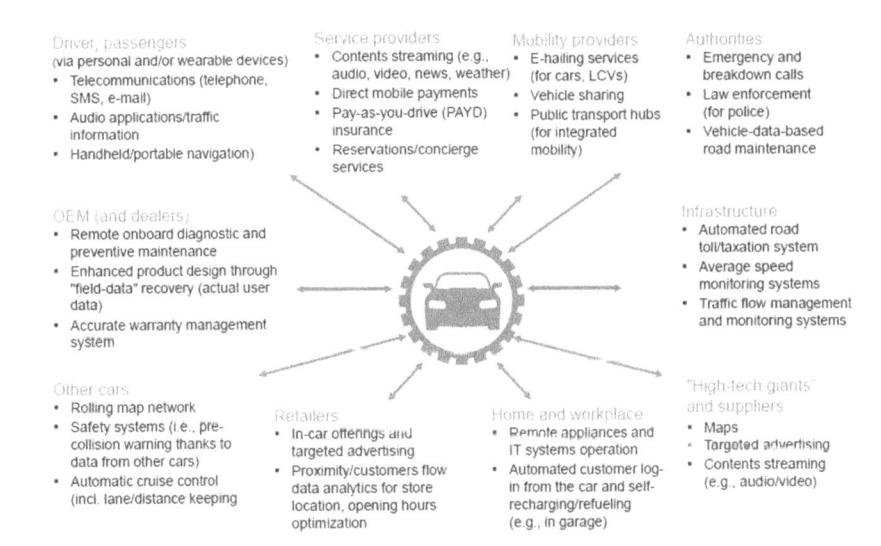

Home and workplace
- Remote appliances and IT systems operation
- Automated customer log-in from the car and self-recharging/refueling (e.g., in garage)

"High-tech giants" and suppliers
- Maps
- Targeted advertising
- Contents streaming (e.g., audio/video)

Fig. 3.6 Car data users/contributors and use case examples. McKinsey & Company II (2016) S.5.

3.2.2 Digital Value Chain and Value Creation Models

In today's business environment, coined by accelerated product innovations and digital customer journeys, companies are facing a variety of market alterations that directly challenge their traditional business models. The digital transformation changes the way companies interact in a B2C and B2B context.[52] Tried-and-tested processes and traditional communication patterns must increasingly become proactive and tailored to the individual requirements to enforce customer retention.[53] The level of digital maturity within the companies thus significantly contributes to achieving added value from the existing information technology tools that can be deployed in the various stages of the value-added model. Business process reengineering and operational excellence thereby create the foundation for a purposeful and efficient digital transformation and will be illustrated in this chapter.

---
[52] cf. BearingPoint (2015) Der Weg zur digitalen Strategie.
[53] cf. Kreutzer, Ralf; Neugebauer, Tim; Pattloch, Annette (2017) Digital Business Leadership S.173.

In this context, a holistic end-to-end approach is required to link internal and external processes along the value chain.[54] All upstream processes must be aligned to match the goal of overall client centricity and the creation of client intelligence[55] that enables a sustainable reorientation of all business functions and foster their efficient contribution to the business objectives (Fig. 3.7).

> The collective term "Industry 4.0" represents the fundamental idea to integrate the industrial production with modern Information and Communication Technology (ICT) to gather and accumulate data from all sources in the value chain and use it to link all process steps. (ROTH 2016 S.69) From advanced forecasting method to smart consumer services, big data-based tools and analytics enable an end-to-end transparency and standardization of the production processes. While the implementation roadmap and further value chain reengineering options are illustrated in detail in Chap. 5, the subsequent section gives a brief insight into the development of a smart digital value chain in the automotive industry. Based on this production optimization, several macro-categories of value creation models are introduced that have an impact on the profitability and cost structure and enable the participation in future platform ecosystem models.

Prerequisite to collecting and processing all data that accrue during the production life cycle is the interlinkage of all data-driven production systems such as Enterprise Resource Planning (ERP) or warehousing systems.[56] The consolidation of information in a digital data center facilitates the usage of big data analytics technologies to forecast future consumer and market behavior based on historical production data and artificial intelligence that recognizes repetitive demand patterns.[57] This enables a pull-oriented production planning and, thus, a customer-centric approach towards the set-up of production plans. Predictive analytic tools aid in forecasting and prioritizing production decisions and manage

---

[54] cf. Siegfried, Patrick (2015) Die Unternehmenserfolgsfaktoren und deren kausale Zusammenhänge S.131–137.
[55] cf. OECD/International Transport Forum (2015) Automated and Autonomous Driving S.10.
[56] cf. PA Consulting Group; The Consumer Goods Forum (2018) AI and Robotics automation in consumer-driven supply chains S.29–31.
[57] cf. Capgemini (2016) Studie IT-Trends 2016 S.18–20.

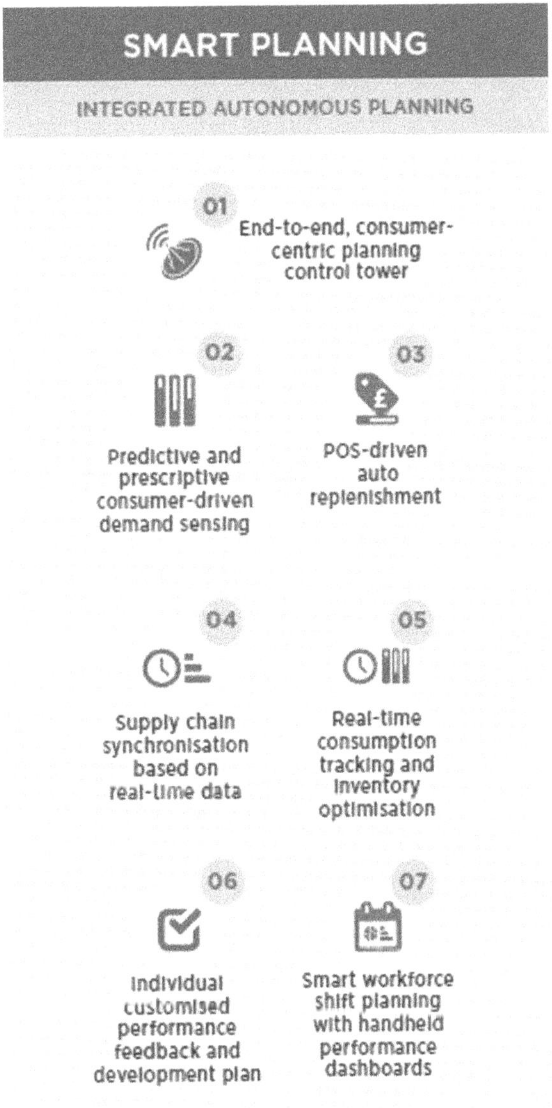

SMART PLANNING
INTEGRATED AUTONOMOUS PLANNING

01 End-to-end, consumer-centric planning control tower

02 Predictive and prescriptive consumer-driven demand sensing

03 POS-driven auto replenishment

04 Supply chain synchronisation based on real-time data

05 Real-time consumption tracking and inventory optimisation

06 Individual customised performance feedback and development plan

07 Smart workforce shift planning with handheld performance dashboards

Fig. 3.7 Potential supply chain automation applications (smart planning). PA Consulting, (2018), S.30

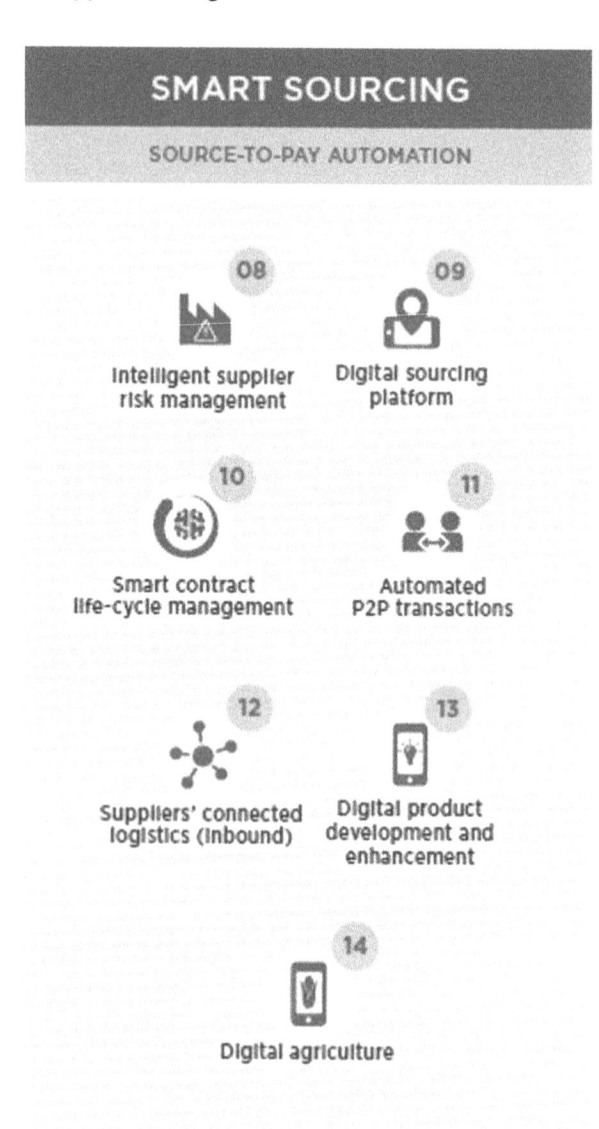

Fig. 3.8 Potential supply chain automation applications (smart sourcing). cf. PA Consulting Group; The Consumer Goods Forum (2018) AI and Robotics automation in consumer-driven supply chains S.30

3 Elements of the Automotive Disruption   41

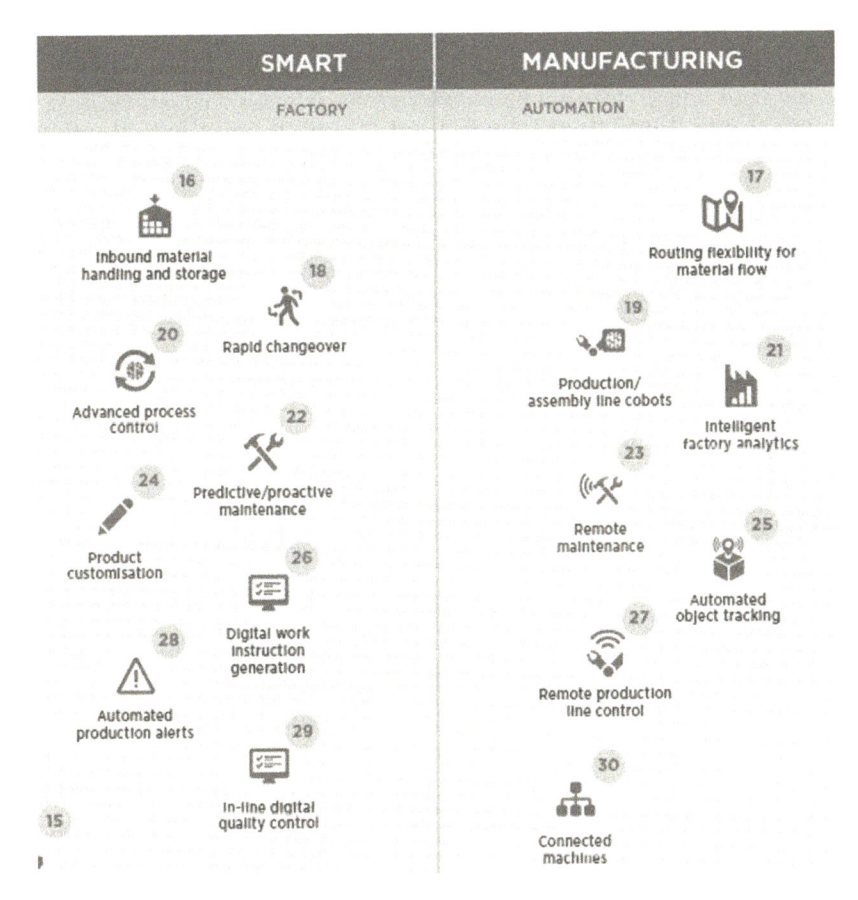

SMART FACTORY	MANUFACTURING AUTOMATION
16 Inbound material handling and storage	17 Routing flexibility for material flow
18 Rapid changeover	19 Production/assembly line cobots
20 Advanced process control	21 Intelligent factory analytics
22 Predictive/proactive maintenance	23 Remote maintenance
24 Product customisation	25 Automated object tracking
26 Digital work instruction generation	27 Remote production line control
28 Automated production alerts	
15 / 29 In-line digital quality control	30 Connected machines

Fig. 3.9 Potential supply chain automation applications (smart manufacturing). PA Consulting, (2018), S.30/31

the synchronization of the asset utilization[58] with replenishment systems that directly record inventory gaps in warehousing. Accordingly, this enables real-time visibility of asset capacities on a product and component level (Figs. 3.8 and 3.9).

Machine learning software, advanced vision systems for scanning and sensing objects, or automated identification using RFID enable real-time adjustments of machine processes, facilitate an early recognition of

---

[58] cf. Mertens, Peter; Barbian, Dina; Baier, Stephan (2017) Digitalisierung und Industrie 4.0 S.46.

> Big data and AI-enabled intelligent analytics are not only shaping the internal value-added chain but also impact on the business relations between automotive manufacturers and their tier-n suppliers. Smart sourcing comprises the realignment of interaction with upstream partners based on digital procurement processes and the creation of a digital interface. This sourcing platform intertwines production and warehousing data of all parties involved and enables automated procurement procedures by creating real-time transparency of commodity and component availability. (OLIVER WYMAN IV 2016) Real-time routing and track-and-trace options simplify modern production forms such as Just-In-Sequence (JIS) and Just-In-Time (JIT). In conclusion, the OEM-supplier-interface improves supplier and transport risk assessment by a transparent and integrated access to mutual planning systems.

> Smart manufacturing models refer to an increased asset and labor capacity utilization. Inbound and outbound logistics transport systems that deploy autonomous vehicles significantly reduce the necessity of human workers. Smart warehousing features such as automated routing that can be adapted using parameters such as production priorities, transport distances and energy efficiency accelerates the production process.

quality defects and allow the introduction of mass customization. An even more advanced IT tool that is progressively applied in all manufacturing-related industry is the development of a virtually-based digital twin, a computer-based reproduction of the production facility, fed with all data systems to conduct sensitivity or scenarios analyses.

The overall IT-driven reorganization and modification of production processes is aligned to match three macro-categories of value creation models.[59] As described above, the combination of analytics and process automation, including tools such as process mining or customer-oriented adjustment of product and services innovation is directly accelerating the generation of revenues. Data-fed research and development investments[60]

---

[59] cf. Blanke, Tobias (2014) Digital Asset Ecosystems S.140.
[60] cf. Heneric, Oliver; Licht, Georg (2005) Europe's Automotive Industry on the Move S.127ff.

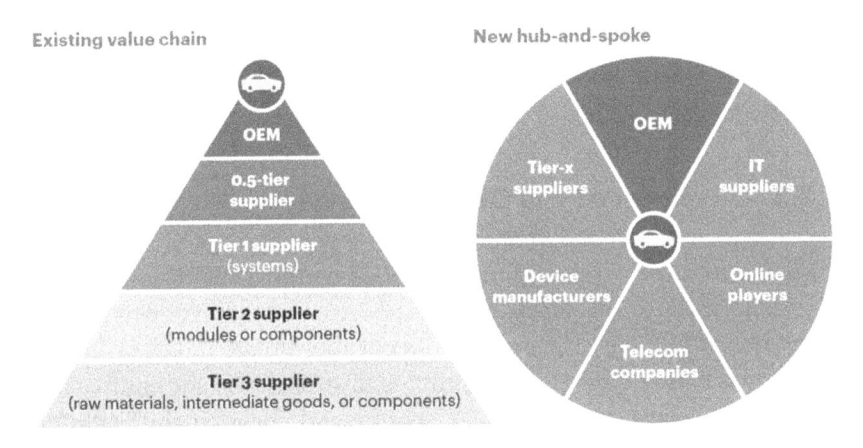

Fig. 3.10 The restructuring of traditional value chains. A.T. Kearney (2016) How Automakers can survive the Self-Driving Era S.13

are reconciled with consumer preferences[61] to create superior customer satisfaction and, thus, foster customer loyalty and retention.

This is accompanied by an overall redesign of the customer journey landscape and the interlinkage with the process roadmap along the digital customer touchpoints. The introduction of an omni-channel management for customer processes increases possibilities for automotive companies to interact with their customers. Since the significant value of Customer Relationship Management (CRM) is continuously increasing with regard to the emergence of platform and ecosystem-based business models,[62] new forms of customer journey creation options will be illustrated in a separate excursus (Fig. 3.10).

Since future revenue streams and profit pools will move from a product-centric to service-centric emphasis, companies are striving to create partnerships and mutual interlinkages via platforms on which customers get access to a diversified spectrum of product and service offerings. All stakeholders from tier-n suppliers to aftersales and additional product service providers are engaging in a comprehensive digital system,

---

[61] cf. Brucker-Kley, Elke; Kykalova, Denisa; Keller, Thomas (2018) Kundennutzen durch Digitale Transformation S. 130–133.
[62] cf. BearingPoint I (2017) BPM macht kurzen Prozess S.6.

so-called ecosystems,[63] that has a central customer interface. This Hub-and-Spoke (HaS) model will be depicted in detail in the subsequent chapter.

3.2.3 Creation of the Digital Ecosystem

Today's automotive industry that is coined by independent manufacturers passes through an alteration towards strengthened cooperation and networking. With an increasing number of automotive and cross-industry companies competing for the connected customer, the automotive landscape has become more complex fragmented. Beyond the traditional business models of manufacturing, distributing, and selling physical goods as well as providing services in the after-sales area, several other business strategies are emerging that are redirecting formerly anchored revenue streams and profit pools.[64] Technology developers, especially data-driven high-tech giants such as Google, Facebook, or Uber as well as new market entrants like Tesla are challenging the incumbent players in their preserve. Start-ups sprouting in different kinds of industries are developing software-based additional services for vehicle users and, thus, creating intellectual property that poses new sources of income beyond the mere selling of automotive hardware. Network orchestrators create and maintain networks of people and information and facilitate interactions and transactions between them.[65]

Especially in Germany, traditional OEMs have a different cultural access to innovation. Automotive products are globally known for high-quality standards, mature technology features, and perfectionist workmanship. Foreign companies, for instance, referring to Tesla, reach shorter time-to-market processes with, in fact, partly half-baked products, but this strategy is creating a more dynamic and innovative corporate culture. Concentrating on operating models prevents German manufacturers from thriving in a digitalized environment that is

---
[63] cf. Bouhai, Nasreddine; Saleh, Imad (2017) Internet of Things S.3.
[64] cf. Pollak, Dale (2017) Like I See it. S.22ff.
[65] cf. Meyer, Gereon; Shaheen, Susan (2017) Disrupting Mobility S.21–23.

incrementally moving from physical products towards less tangible services. New business models are asset light and underpinned by customer intelligence and use partner ecosystems for co-innovation and investment.[66] The triumvirate of operating systems, applications, and hardware respectively vehicles are depicting the corner pillars of novel business models. Industry experts predict a future shift in the way companies are competing. Since the platforms and services are not subject to standardization, multiple concepts will emerge with different stakeholders flocking together. The competition will thus not take place between single companies but between different platforms and their members.[67] In such an ecosystem, OEMs, suppliers, and cross-industry players cooperate by

	Reciprocal	Bundled
Centralized	Supply Systems	Platforms
Decentralized	Common Destiny	Expanding Communities

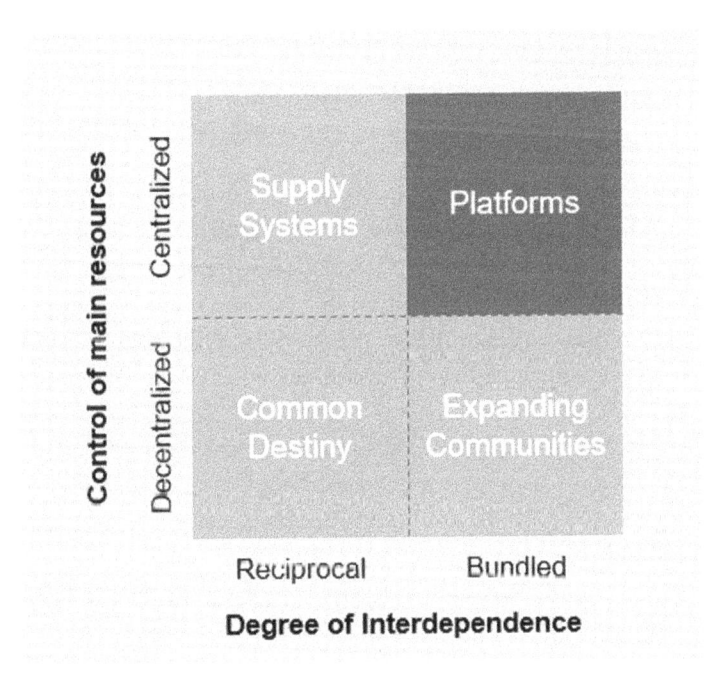

Control of main resources / Degree of Interdependence

Fig. 3.11 Four distinct types of ecosystems. Author's own representation based on BearingPoint/IIHD Institute, (2017), S.8

---

[66] cf. Herrmann, Andreas; Brenner, Walter; Stadler, Rupert (2018) Autonomous Driving S.89–92.
[67] cf. A.T. Kearney (2016), S.15.

using the platform to be able to reach a critical size without vast investments or resource input.

These digital, platform-based business models are more resilient than traditional models since the network community disposes of better customer insight, creates economies of scale from network effects, and benefits from very low marginal costs of growth due to its mere software-based network model.[68] German automotive producers must incorporate these digital-platform-based business models into the center of their strategy while driving synergies between their traditional and novel business activities (Fig. 3.11).

> There are different types of ecosystems dependent on the criteria control of central resources and degree of cross-integration. Supply systems depict a centralized ecosystem in which one company steers a network of partners that contribute to the business activities. Primary interest is the secured access to resources on upstream value-added stages.

Typical phenotypes are supplying partnerships between OEMs and tier-n suppliers. Expanding communities, in contrast, consist of many co-equal participants that are focused on a central, but unrestricted resource. Access to this non-proprietary resource is the essential criterion of differentiation regarding other ecosystems. Knowledge-intense communities, for instance in the computer software solutions area, are characteristic examples.

The platform model which is progressively gaining significance for the automotive sector is based on one pivotal company functioning as designer and steering authority providing a central resource to the participants of the platform.[69]

The main unique feature of this model is the multiplication of single, independent business activities and services within the platform. Participating partner companies provide their own products and services and, thus, mutually influence the coverage and range of activities.

---

[68] cf. Colugnati, Fernando; Lopes, Lia Carrari (2010) Digital Ecosystems S.31.
[69] cf. PwC; NTT Data Deutschland (2013) Automotive Retail S.34.

Well-known examples are eBay, Amazon, or Facebook that provide an open platform for sellers, purchasers, and content creators.

The prevailing hyporesearch work of the future competitive structure emanates from the idea that companies will no longer be defined by their share in the value-added process[70] or their range of business activities. Competition will take place between networks of companies, that cover all stages in the value-added chain up to the customer. The ecosystem, in this context, is configured as a plug-and-play network. Based on the principle of modularity, new participants and their product portfolios can be integrated to strengthen the network's value proposition. There is a large spectrum of competitive advantages that result from the sharing character of this model. Industry boundaries are becoming blurred and the physical possession of resources is subordinate.[71] By this decoupling of individual possession on grounds of a shared resource basis, new sources of value-added streams and product offerings are unlocked. Partners

Fig. 3.12 Network effects exemplified with Uber. Author's own representation based on BearingPoint/IIHD Institute, (2017), S.14

---

[70] cf. Bundschuh, Dominik (2017) Industrie 4.0 in Deutschland S.38.
[71] cf. Aldrich, Douglas F. (1999) Mastering the Digital Marketplace S.91.

benefit from the complementary concept of resource bundling and can extend their product and service portfolio based on mutual access possibilities. This reciprocal interlinkage leads to a more balanced and diversified risk management since all partners can benefit from each other's contributions and can focus on their own core competences without losing control of the comprehensive data flows. Furthermore, the direct customer approach, triggered by a free access to the platform by all users, cuts off intermediaries which lowers prices and gives the customer unrestricted autonomy in the selection of suited products and services.[72] In conclusion, this creates added value for the customer. Complex quality and control mechanisms are replaced by using community feedback to identify inefficient or business impairing processes.[73] The integration of the users into the ongoing redesign and improvement of the platform will further increase customer satisfaction and perception of affiliation. In the graphic below, the general concept of platform growth opportunities is illustrated, exemplified with Uber (Fig. 3.12).

Network effects are the major driver of the successful platform and ecosystem concepts. In the Uber example, more participating drivers lead to a higher geographical coverage, in turn increasing the attractiveness and availability of the services for more customers. The gradually rising number of drivers and users directly shorten idle and queue times, improving capacity usage and lowering customer prices.[74] In conclusion, automotive companies have the opportunity to create a self-controlled platform that combines the sales of vehicles with extended products and services provided by a partnership community. This concept improves existing customer relations by providing a novel digital touchpoint and attracts new customer segments.[75] The customer insight companies receive by analyzing and processing customer activities can be used to improve product and service offerings and, thus, reinforce revenue streams.

---

[72] cf. KPMG I (2017) Reimagine places: Mobility as a Service S.13.
[73] cf. BearingPoint; IIHD Institut (2017) Ecosysteme & Plattformen verändern die Handelslandschaft S.8–10.
[74] cf. KPMG I (2016) I See. I Think. I Drive. (I Learn) S.21.
[75] cf. World Economic Forum (2016) Digital Transformation of Industries S.19.

References

Aldrich, D. F. (1999). *Mastering the digital marketplace. Practical strategies for competitiveness in the new economy* (1. Aufl.). John Wiley & Sons.

Anderson, J., Kalra, N., Stanley, K., Sorenson, P., Samaras, C., & Oluwatola, O. (2016). *Autonomous vehicle technology. A guide for policymakers.* Rand Corporation.

BearingPoint. (2015). *Der Weg zur digitalen Strategie*, S. 1–2. Retrieved April 4, 2021, from https://www.bearingpoint.com/files/BEDE15_0969_DE_Digitale_Strategien_final.pdf&download=0&itemId=133117

BearingPoint I. (2017). *BPM macht kurzen Prozess. Business Process Management Studie 2017*, S. 1–12. Retrieved March 16, 2021, from https://www.bearingpoint.com/files/BearingPoint_Management_Summary_BPM_Studie_2017.pdf&download=1&itemId=442094

BearingPoint III. (2017). *Roboter, Rebellen, Relikte. Überkommene Strukturen behindern die Digitale Transformation*, S. 1–21. Retrieved May 11, 2021, from https://www.bearingpoint.com/files/Digitalisierungsmonitor_2017.pdf&download=1&itemId=470351

BearingPoint, IIHD Institut für internationales Handels- und Distributionsmanagement. (2017). Ecosysteme & Plattformen verändern die Handelslandschaft. Wie branchenübergreifende Koopertionen die Wettbewerbsstrukturen und -logiken des Handels von morgen bestimmen. In *Retail & Consumer* (No. 12), S. 1–24. Retrieved April 18, 2021, from https://www.bearingpoint.com/files/BEDE17_1148_RP_DE_Ecosysteme_und_Plattformen_ver%C3%A4ndern_die_Handelslandschaft.pdf&download=1&itemId=461481

Blanke, T. (2014). *Digital asset ecosystems. Rethinking crowds and clouds.* Chandos Publishing.

Bock, A. J., & George, G. (2017). *The business model book. Design, build and adapt business ideas that driver business growth* (3. Aufl.) Pearson Education Limited.

Bouhai, N., & Saleh, I. (2017). *Internet of things. Evolutions and innovations.* ISTE Ltd. (Digital Tools and Uses Set, 4).

Brucker-Kley, E., Kykalova, D., & Keller, T. (2018). *Kundennutzen durch Digitale Transformation. Business-Process-Management-Studie - Status Quo und Erfolgsmuster.* Springer Gabler.

Bundesverband Digitale Wirtschaft. (2016). *Connected Cars - Geschäftsmodelle*, S. 2–15. Retrieved May 11, 2021, from https://www.bvdw.org/fileadmin/

bvdw/upload/publikationen/digitale_transformation/Diskussionspapier_Connected_Cars_Geschaeftsmodelle.pdf

Bundschuh, D. (2017). *Industrie 4.0 in Deutschland. Der digitale Wandel in der Automobilindustrie*. GRIN Verlag, Open Publishing GmbH.

Capgemini. (2016). *Studie IT-Trends 2016. Digitalisierung ohne Innovation?* S. 3–46. Retrieved April 4, 2021, from https://www.capgemini.com/de-de/wp-content/uploads/sites/5/2016/02/it-trends-studie-2016.pdf

Capgemini Consulting. (2017). *Beyond the car*, S. 1–36. Retrieved February 25, 2021, from https://www.capgemini.com/consulting-de/wp-content/uploads/sites/32/2017/05/cars-online-study-2017.pdf

Colugnati, F., & Lopes, L. C. (2010). Digital ecosystems. *Third international conference, OPAALS 2010*. Springer Verlag.

Deloitte I. (2017). *The future of the automotive value chain. 2025 and beyond*, S. 1–64. Retrieved April 4, 2021, from https://www2.deloitte.com/content/dam/Deloitte/us/Documents/consumer-business/us-auto-the-future-of-the-automotive-value-chain.pdf

Deloitte University Press. (2015). *Patterns of Disruption. Anticipating disruptive strategies in a world of unicorns, black swans and exponentials*, S. 1–34. Retrieved May 11, 2021, from https://www2.deloitte.com/content/dam/Deloitte/br/Documents/technology/Patterns-of-disruption.pdf

Deloitte University Press. (2017). *Governing the future of mobility. Opportunities for the US government to shape the new mobility ecosystem*, S. 1–26. Retrieved February 25, 2021, from https://www2.deloitte.com/content/dam/Deloitte/nl/Documents/consumer-business/deloitte-nl-fom-how-to-govern-the-future-of-mobility.pdf

Deutsche Bank AG. (2017). The digital car. More revenue, more competition, more cooperation. In *German Monitor—The digital economy and structural transformation*, S. 1–36. Retrieved March 16, 2021, from https://www.dbresearch.com/PROD/RPS_EN-PROD/PROD0000000000446248/The_digital_car%3A_More_revenue%2C_more_competition%2C_m.pdf

Elgar, E. (2007). *The digital business ecosystem*. Edwar Elgar Publishing Limited.

ESCP Europe. (2017). *Automotive disruption switch to digital services*, S. 1–37. Retrieved March 9, 2021, from https://www.eurogroupconsulting.com/sites/eurogroupconsulting.fr/files/document_pdf/etude_eurogroup_escp_europe_-_digital_services_in_automotive_distributio.pdf

European Commission. (2017). *Digital transformation monitor. Autonomous cars: A big opportunity for European industry*, S. 1–6. Retrieved March 3, 2021, from https://ec.europa.eu/growth/tools-databases/dem/monitor/sites/default/files/DTM_Autonomous_cars v1_1.pdf

EY. (2017). *The evolution in self-driving vehicles. Trends and implications for the insurance industry*, S. 1–12. Retrieved April 18, 2021, from https://www.ey.com/Publication/vwLUAssets/ey-self-driving-vehicle-v2/$FILE/ey-self-driving-vehicle-v2.pdf

Federal Ministry of Transport and Digital Infrastructure. (2015). *Strategy for automated and connected driving. Remain a lead provider, become a lead market, introduce regular operations*, S. 1–32. Retrieved February 25, 2021, from https://www.bmvi.de/SharedDocs/EN/publications/strategy-for-automated-and-connected-driving.pdf?__blob=publicationFile

Friess, P. (2014). *Internet of things. From research and innovation to market development*. River Publishers.

Hencric, O., & Licht, G. (2005). *Europe's automotive industry on the move. Competitiveness in a changing world*. Physica Verlag (32).

Herrmann, A., Brenner, W., & Stadler, R. (2018). *Autonomous driving. How will the driverless revolution change the world*. Emerald Publishing Ltd.

Hülsmann, M., & Fornahl, D. (2014). *Evolutionary paths towards the mobility patterns of the future*. Springer Verlag.

Jurgen, R. K. (2013). *Autonomous vehicles for safer driving*. SAE International (Progress in technology series, 158).

Karls, I., & Mueck, M. (2018). *Networking vehicles to everything. evolving automotive solutions*. Walter de Gruyter GmbH.

Kearney, A. T. (2016). *How automakers can survive the self-driving era. A.T. Kearney study reveals new insights on who will take the pole position in the $560 billion autonomous driving race*, S. 1–36. Retrieved March 16, 2021, from https://www.kearney.com/automotive/article?/a/how-automakers-can-survive-the-self-driving-era

KPMG I. (2016). *I see. I think. I drive. (I learn). How Deep Learning is revolutionizing the way we interact with our cars*, S. 1–44. Retrieved March 16, 2021, from https://assets.kpmg.com/content/dam/kpmg/se/pdf/komm/2016/se-isee-ithink-idrive-ilearn.pdf

KPMG I. (2017). *Reimagine places: Mobility as a service. The mobility as a service (MaaS) requirements index 2017*, S. 1–32. Retrieved July 2, 2021, from https://assets.kpmg.com/content/dam/kpmg/uk/pdf/2017/08/reimagine_places_maas.pdf

KPMG II. (2017). *Global automotive executive survey 2017*, S. 1–56. Retrieved May 6, 2021, from https://assets.kpmg.com/content/dam/kpmg/es/pdf/2017/global-automotive-executive-survey-2017.pdf

Kreutzer, R., Neugebauer, T., & Pattloch, A. (2017). *Digital business leadership. digital transformation, business model innovation, agile organization, change management.* Springer Fachmedien GmbH.

Linnhoff-Popien, C., & Schneider, R. (2018). *Digital marketplace unleashed.* Springer Verlag GmbH.

McKinsey & Company II. (2015). *Urban mobility at a tipping point*, S. 3–22. Retrieved May 6, 2021, from https://www.mckinsey.com/business-functions/sustainability-and-resource-productivity/our-insights/urban-mobility-at-a-tipping-point

McKinsey & Company II. (2016). *Car data: paving the way to value-creating mobility. Perspectives on a new automotive business model*, S. 5–21. Retrieved June 2, 2021, from https://www.mckinsey.de/files/mckinsey_car_data_march_2016.pdf

McKinsey & Company III. (2016). Monetizing car data. New service business opportunities to create new customer benefits. In *Advanced Industries*, S. 3–57. Retrieved February 22, 2021, from https://www.mckinsey.com/~/media/McKinsey/Industries/Automotive%20and%20Assembly/Our%20Insights/Monetizing%20car%20data/Monetizing-car-data.ashx

Mertens, P., Barbian, D., & Baier, S. (2017). *Digitalisierung und Industrie 4.0 - eine Relativierung.* Springer Fachmedien GmbH.

Meyer, G., & Shaheen, S. (2017). *Disrupting mobility. Impacts of sharing economy and innovative transportation on cities.* Springer International Publishing AG.

Müller, S. (2016). *Internet of things (IoT). Ein Wegweiser durch das Internet der Dinge.*

Neckermann, L. (2015). *The mobility revolution.* Troubador Publishing Ltd.

Nieuwenhuis, P., & Wells, P. (2015). *The global automotive industry* (1. Aufl.). John Wiley & Sons, Ltd.

OECD/International Transport Forum. (2015). *Automated and autonomous driving. Regulation under uncertainty*, S. 1–32. Retrieved April 27, 2021, from https://www.itf-oecd.org/sites/default/files/docs/15cpb_autonomous-driving.pdf

PA Consulting Group, The Consumer Goods Forum. (2018). *AI and Robotics automation in consumer-driven supply chains. A rapidly evolving source of competitive advantage*, S. 1–54. Retrieved June 6, 2021, from https://www.theconsumergoodsforum.com/wp-content/uploads/2018/04/201805-CGF-AI-Robotics-Report-with-PA-Consulting.pdf

Pollak, D. (2017). *Like i see it. Obstacles and opportunities shaping the future of retail automotive.* vAuto Press.

PWC, NTT Data Deutschland. (2013). *Automotive Retail—Die Zukunft beginnt jetzt*, S. 12–56. Retrieved June 6, 2021, from https://www.pwc-wissen.de/pwc/de/shop/publikationen/Automotive+Retail+-+Die+Zukunft+beginnt+jetzt!/?card=13001

PWC - Strategy&. (2016). *Connected car report 2016. Opportunities, risk, and turmoil on the road to autonomous vehicles*, S. 5–63. Retrieved April 27, 2021, from https://www.strategyand.pwc.com/reports/connected-car-2016-study

Roland Berger. (2017). *Automotive disruption radar. Tracking disruption signals in the automotive industry*, S. 1–20. Retrieved July 17, 2021, from https://www.rolandberger.com/publications/publication_pdf/roland_berger_disruption_radar.pdf

Siegfried, P. (2014). *Knowledge transfer in service research—Service engineering in startup companies* EUL-Verlag, Siegburg.

Siegfried, P. (2015). *Die Unternehmenserfolgsfaktoren und deren kausale Zusammenhänge, Zeitschrift Ideen- und Innovationsmanagement.* Deutsches Institut für Betriebswirtschaft GmbH/Erich Schmidt Verlag.

Siegfried, P. (2021). *Enterprise management automobile industry business cases.* BoD.

The Boston Consulting Group I. (2016). Digital reroutes the auto purchase journey and OEM strategies. In *bcg.perspectives*, S. 1–6. Retrieved June 6, 2021, from https://www.bcg.com/publications/2016/automotive-retail-digital-reroutes-auto-purchase-journey-oem-strategies.aspx

U.S. Department of Commerce, Economics and Statistics Administration. (2017). The employment impact of autonomous vehicles. In *ESA Issue Brief* (5), S. 1–33. Retrieved March 12, 2021, from http://www.esa.doc.gov/sites/default/files/Employment_Impact Autonomous Vehicles_0.pdf

UBS. (2017). *Longer term investments. Smart mobility*, S. 1–25. Retrieved March 17, 2021, from https://www.ubs.com/content/dam/WealthManagement Americas/documents/smart-mobility.pdf

Wedeniwski, S. (2015). *The mobility revolution in the automotive industry. How not to miss the digital turnpike.* Springer Verlag GmbH.

Winkelhake, U. (2017). *Die digitale Transformation der Automobilindustrie. Treiber - Roadmap -Praxis.* Springer Verlag GmbH.

World Economic Forum. (2016). Digital transformation of industries. Automotive industry. In *Digital enterprise*, S. 1–45. Retrieved May 11, 2021, from http://reports.weforum.org/digital-transformation/wp-content/blogs.dir/94/mp/files/pages/files/digital-enterprise-narrative-final-january-2016.pdf

Wyman, O. (2017). *The Oliver Wyman automotive manager*, S. 1–61. Retrieved April 23, 2021, from www.oliverwyman.de/content/dam/oliver-wyman/v2/publications/2017/jun/OliverWyman_AutomotiveManager2017_web.pdf

4

Urban Mobility Revolution: A Quantitative Analysis

4.1 Predicted Customer Demand Patterns

The survey that will be presented in this chapter was conducted within the framework of this research work, providing insight into customers' acceptance and potential usage behavior of connectivity features and autonomous vehicles. Hundred participants were asked to indicate their personal opinion and attitude towards current and prospective mobility solutions. The pie charts compiled below, give an abstract on the composition of the survey group. While the gender affiliation is well-balanced, respondents under the age of 26 clearly represent the majority, accounting for a 56% share, whereas the age cluster between 26 and 45 constitutes one-third of the total number of participants. 44 of 100 respondents reside in large metropolitan areas with more than 500,000 citizens while the remaining survey group is nearly equally allocated to the other quantity clusters ranging from less than 20,000 to up to 500,000 inhabitants. As the distribution of age hypothesizes, 57% of the respondents earn less than 3000 Euros of gross income per month, average income levels between 3000 and 6000 Euros account for 31% of the survey participants while only 12 of 100 earn more than 6000 (Fig. 4.1).

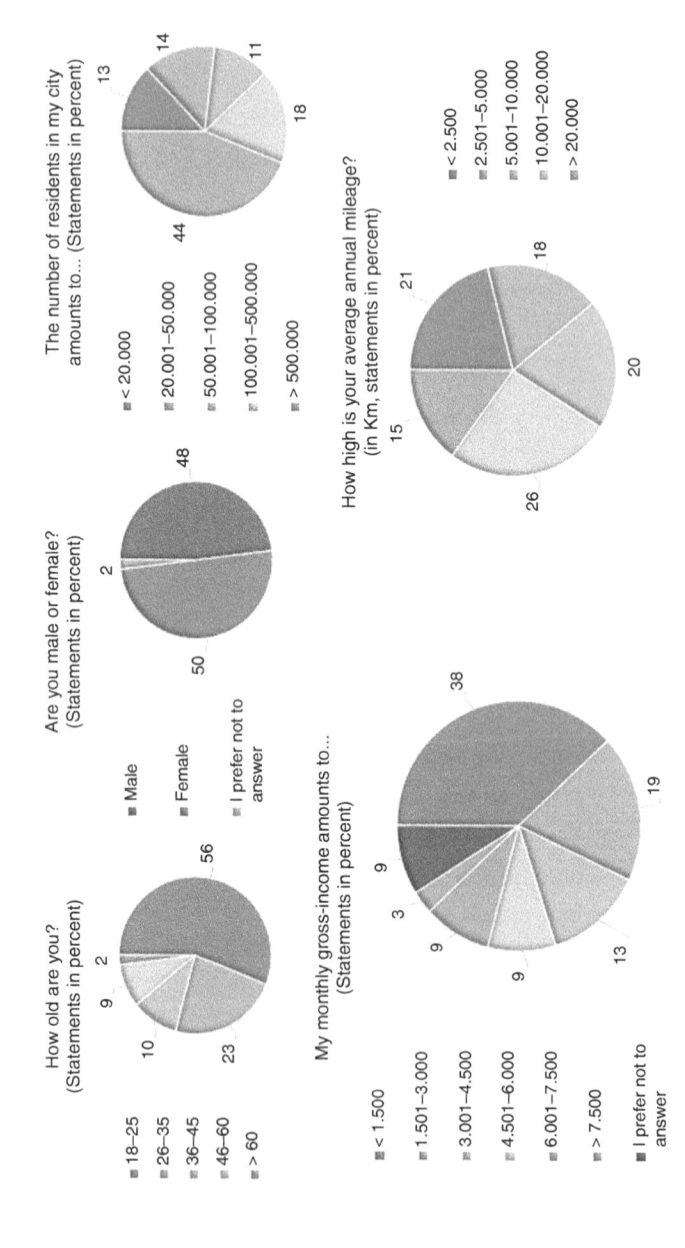

Fig. 4.1 Demographic characteristics of the survey group. Author's own representation

With respect to the average annual mileage, the shares of responses are uniformly distributed with a marginal relative majority of the category ranging from 10,001 to 20,000 kilometers per year.

Before continuing with an in-depth assessment of prospective consumer segments, an identification of the Value of Time (VoT), induced by self-driving vehicles (SDV), and a determination of the forecasted market volume, potential biases have to be considered that result from the statistical distribution of survey participants.

The monthly gross income, city size, and age of participants, in particular, deviate evidently from the average value of the German population.[1] This affects the likelihood of answers being unilaterally biased towards a higher prioritization of cost-related and price-sensitive response options. The dislocation of the normal distribution may, however, be compensated for by the increased interest of younger participants in novel features and innovations. Above-average tech affinity and juvenile curiosity are typically characteristics attributed to younger consumer segments.[2] Another indication that fortifies this research work is the high share of respondents residing in densely populated cities. Autonomous driving features promise a relief from congested and hectic city traffic, thus depicting a higher value proposition for participants affected by bothersome driving conditions.[3] In addition, the number of urban dwellers not owning private vehicles exceeds the number of those living in rural areas. Hence, many could be interested in flexible, on-demand, and shared mobility models making private car ownership dispensable.[4] To what extent these conjectures apply, will be illustrated in the subsequent evaluation.

Beyond investing in technological advances in the hardware and software field including sensor systems, human–machine interfaces, artificial intelligence, or big data processing tools,[5] OEMs must not neglect the demand side of this equation. Innovations typically hold an intrinsic risk

---

[1] cf. PwC; NTT Data Deutschland (2013) Automotive Retail S.21–22.
[2] cf. IBM Institute for Business Value (2010) A new relationship - people and cars S.7–9.
[3] cf. Anderson, James et al. (2016) Autonomous Vehicle Technology S.16–18.
[4] cf. Deloitte (2016) Autonomes Fahren in Deutschland S.9–12.
[5] cf. Herrmann, Andreas; Brenner, Walter; Stadler, Rupert (2018) Autonomous Driving S.121.

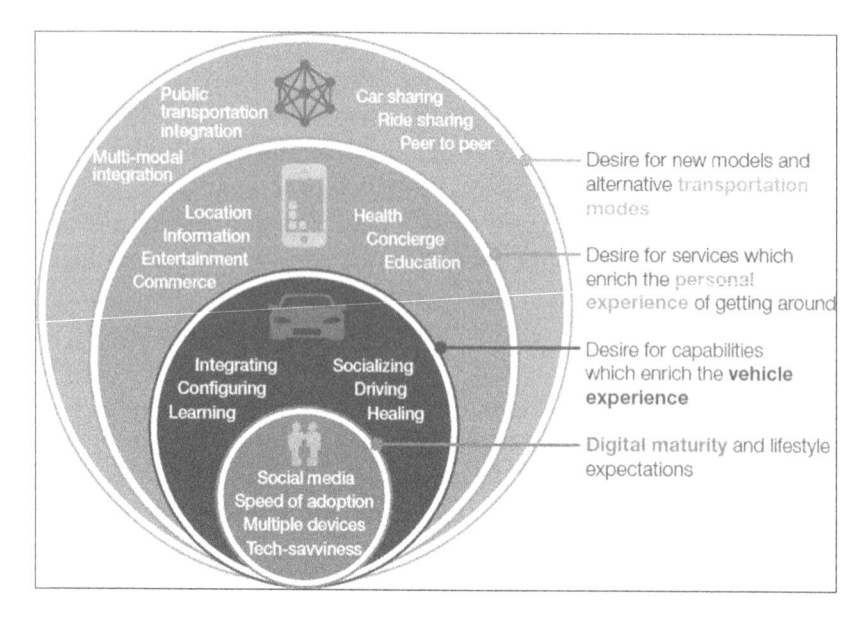

Fig. 4.2 Digital mobility interest levels and the expectations for future mobility scenarios

of encountering resistance in the initial stages of their market introduction. In order to avoid teething troubles, the automotive industry should further vivify the public discussion to attune future customers to the altering mobility and AV landscape. Engaging consumers on the technology development path will contribute to a more frictionless and eagerly anticipated market launch.[6] To identify potential incubators for an accelerated market penetration, customers can be clustered by their digital maturity respectively their interest in digital mobility (Fig. 4.2).

Consumers who consider themselves tech-savvy, early adopters of the latest technologies, and are using social media or similar internet platforms more frequently, typically show higher expectations for novel digital innovations in the vehicle and mobility services area.[7] A more in-depth appraisal of potential customers provides an insight into the mobility preferences and technology affinity based on which four distinct

---
[6] cf. Skilton, Mark (2016) Building Digital Ecosystem Architectures S.62.
[7] cf. KPMG I (2018) Autonomous Vehicle Readiness Index S.49.

consumer groups can be identified. As depicted in the figure above, the different clusters are collocated in a valence shell scheme that illustrates the extent of the digital mobility interest of the corresponding group.

OEMs need to comprehend differing customer expectations towards product and service specifications dependent on their individual proclivities and lifestyles to accurately match their consumers with digital technologies within their portfolio.[8] Coming back to the factors determining a successful progression in innovation commercialization, customers can be clustered into four types showing different attitudes towards novel technology adoption. The so-called "pacesetters" represent early adopters that are eager to try new mobility services and are decisive for the initial speed of market diffusion. "Fast-followers" are observing customers pioneering in product testing and quickly succeed the pace-setting group.[9] These two segments combined account for roughly 48% of the respondents surveyed by IBM in 2016.[10] Consumers who view technology conservatively but tend to adopt innovations as soon as their value has been established publicly, represent the relative majority of consumers with 38%. "Spectators" are usually inflexible about exploring new mobility solutions, indicate neglectable affinity to technological features or devices, and intend to retain the status quo. In conclusion, OEMs need to stimulate the demands set by pacesetters and fast-followers to increase the marketability of new mobility models and connectivity features.[11]

The aforementioned ratio of customers interested in testing innovations, particularly autonomous vehicles, is confirmed by the survey conducted within the context of this research work. Two-thirds of the participants stated that they would likely or very likely use self-driving cars, while 40% were fairly reluctant (Fig. 4.3).

---

[8] cf. BearingPoint II (2017) Mensch & Maschine im Kundenservice.
[9] cf. McKinsey & Company I (2015) Ten ways autonomous driving could redefine the automotive world S.3.
[10] cf. IBM Institute for Business Value (2016) A new relationship – people and cars S.7.
[11] cf. McKinsey & Company I (2015) Ten ways autonomous driving could redefine the automotive world S.3.

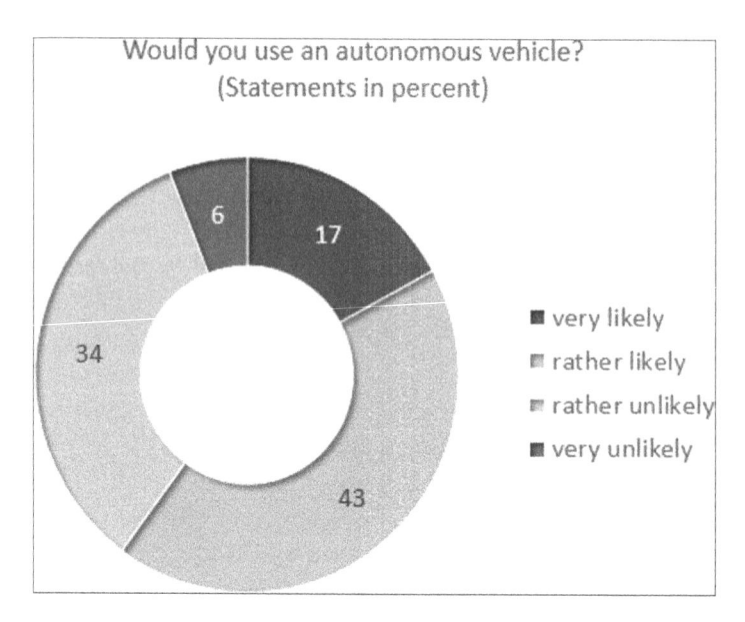

Fig. 4.3 Would you use an autonomous vehicle? Source: Author's own representation

> A more detailed percental allocation of responses according to demographic characteristics and mobility habits revealed that the appeal to try novel autonomous technology increases in negative correlation with age and city size. Hence, young urbanites are distinctly more open-minded than old ruralists.

Long-term commuters and respondents with high private driving times, can be assigned to this category of customers, as well, eagerly anticipating the introduction of autonomous vehicles to gain travel convenience. Coherently, these characteristic groups have a higher willingness-to-pay for autonomous functions, especially those which support the driver in city and highway traffic.[12] Automotive manufactures seeking to convince reluctant customers have several options in this regard. Users fearing technical defects and doubting the maturity of the automated features could be persuaded by a demonstration of the func-

---

[12] cf. Humboldt-Universität zu Berlin (2016) User Perspectives on Autonomous Driving S.45ff.

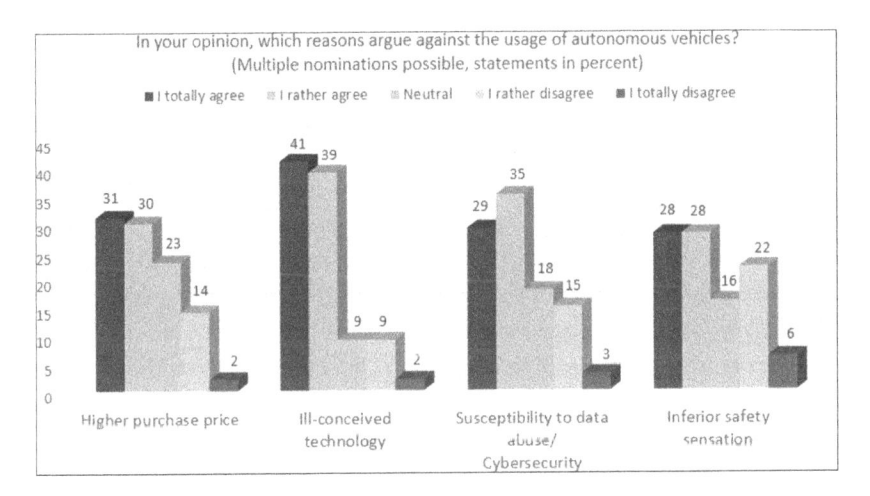

4 Urban Mobility Revolution: A Quantitative Analysis   61

Fig. 4.4 In your opinion, which reasons argue against the usage of autonomous vehicles? Source: Author's own representation

tionalities under realistic traffic conditions.[13] Fifty-nine percent stated that OEMs could incentivize a purchase by assuming liability for vehicle-inflicted incidents that occur within the autonomous driving mode. A statistically proven safety level that guarantees lower frequencies of accidents or casualties would be a convincing argument for 51% of the respondents. Forty-three percent require more explanation and diagrams that illustrate the technical mode of operation of autonomous functions and roughly one quarter could be convinced by cost savings within the context of privileged insurance tariffs (Fig. 4.4).

In the ranking of reasons that argue against the usage of autonomous vehicles, ill-conceived technologies and their susceptibility to data abuse as well as concerns regarding weak points in the cybersecurity systems are rated slightly more decisive than an inferior safety sensation and higher purchase prices with percental shares of approval ranging from 80% to 56%. Further, frequently cited arguments for customers' apprehension toward SDVs include fear of loss of control,[14] defective components, and little knowledge about the functionality as described in a survey

---

[13] cf. Deloitte III (2017) What's ahead for fully autonomous driving S.3–7.
[14] cf. EY II (2016) How much human do we need in a car?.

conducted by the Boston Consulting Group in 2016.[15] Beyond safety concerns, one-third of the respondents stated that driving is a pleasure to them, thus derogating their willingness to pay extra for self-driving functionalities.

In the prioritization of reasons that argue for a usage of autonomous driving functions, respondents agree on the majority about the two most determining advantages. Improved driving convenience and more idle time during travel are both quoted by 78% of the survey group. Participants seemed to be less convinced by a possible reduction of environmental pollution or increased traffic safety. Other studies confirm these results.[16] The preponderate share of arguments referred to autonomous functions as a relief in congested traffic or bothersome driving tasks such as a search for parking spots or refueling. As passengers are gaining idle time, 40% of the respondents see an opportunity to multitask and be productive while driving (Fig. 4.5).

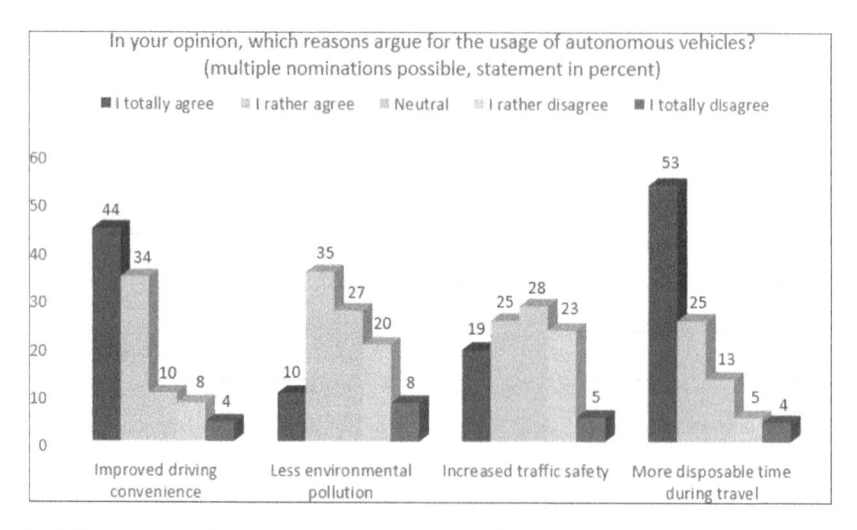

Fig. 4.5 In your opinion, which reasons argue for the usage of autonomous vehicles? Source: Author's own representation

---

[15] cf. The Boston Consulting Group; World Economic Forum (2016) Self-driving vehicles, robot-taxis, and the urban mobility revolution S.8–9.

[16] cf. McKinsey & Company III (2015) Wettlauf um den vernetzten Kunden S.19–21.

In conclusion, to achieve a frictionless market introduction, OEMs need to consider two decisive determinants. Autonomous driving features and connected services should be configured in a user-oriented way, tailored to specific requirements that are assigned to different potential customer segments setting the pace for the market diffusion.[17] A constructive participation of automotive manufacturers in the public discussion about advances in the field of autonomous innovations may further contribute to steering consumers' perception of the technology's maturity and safety level. Customers need to fully understand the value proposition that self-driving vehicles are promoting.

The value of time, as will be illustrated in the subsequent sub-chapter, represents an additional key driver to convince consumers of the benefits coming from AVs.[18] As connectivity is becoming increasingly commoditized, even in models and derivatives that are ascribed to small and compact-sized vehicle classes, connected features are considered a critical purchasing factor for the majority of car owners and prospective customers.[19] Recent studies conducted by McKinsey revealed that 28% of new-car buyers prioritize connectivity over traditionally decisive criteria such

Purchase Criteria	%
Connectivity Features	43
Sustainability/Fuel efficiency	47
Design & Brand reputation	59
Safety/Reliability	71
Price-performance ratio	72

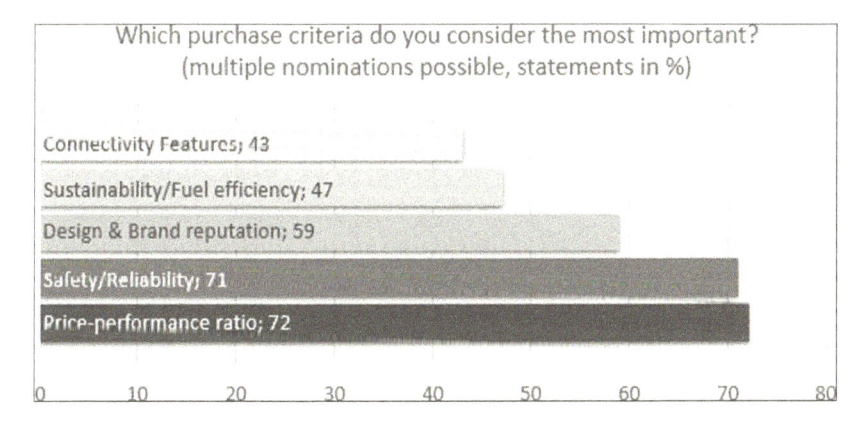

Fig. 4.6 Which purchase criteria do you consider the most important? Source: Author's own representation

---

[17] cf. Kreutzer, Ralf; Neugebauer, Tim; Pattloch, Annette (2017) Digital Business Leadership S.127.
[18] cf. Blanke, Tobias (2014) Digital Asset Ecosystems S.21.
[19] cf. Capgemini (2016) Studie IT-Trends 2016 S.18–19.

as engine power or fuel efficiency and 13% even consider the non-integration of the car into the IoT as a criterion for exclusion (Fig. 4.6).[20]

The survey results, assessed within the study for this research work, are not entirely congruent since the relevance for connectivity features is only ranked in fifth place, but this can mainly be attributed to the composition of the survey group. Fifty-six percent of the participants are younger than 26 years while 57%, most of which are part of the aforementioned age group, earn less than 3000 Euros of gross income per month. Hence, cost-referred purchase criteria such as price-performance ratio or fuel efficiency hold higher significant value for this population group. Nevertheless, key findings have shown that especially younger consumers within the age spectrum of 18–39 are most willing to switch their automotive manufacturer for connected features and are most interested in owning autonomous vehicles with 86.5% compared to an average value of 76%. In general, there is a clear tendency of elderly respondents to stick to their familiar car brand and to prefer conventional driving. In comparison with the Chinese and US markets, German customers are more reluctant to switch manufacturers for gaining access to data and media applications and moreover to pay for connected services in a subscription-based model.[21]

This phenomenon can partly be ascribed to the average age of the populations which diverge by roughly 9 years compared to Germany, thus entailing an age-related higher affinity to new technologies.[22] Another study conducted by McKinsey & Company in 2014, has identified and clustered five different customer segments and their corresponding preferences and attitudes towards car connectivity. These distinct groups of new-car buyers can further be sub-divided into two peer groups that differ regarding the significant value both assign to connected features. "Purists/minimalists" and "price-conscious traditionalists" are typically not affine to novel technology innovations or devices and express their concerns about data privacy and protection in a greater extent. Beyond their disinterest in car connectivity, these consumer groups spend less

---

[20] cf. McKinsey & Company III (2016) Monetizing car data S.14–16.
[21] cf. Roland Berger; fka Forschungsgesellschaft (2016) Index "Automatisierte Fahrzeuge"S.12–18.
[22] cf. PwC; NTT Data Deutschland (2013) Automotive Retail S.21–22.

time per week driving and show a high price sensitivity with respect to the average initial purchase costs. The second peer group comprising "maxed-out car enthusiasts," "integration and entertainment lovers" as well as "safe and secure navigators" represents a cluster that is more appealing to OEMs and third-party providers seeking to commercialize connectivity features and digital services (Fig. 4.7).

These potential buyers appreciate advanced features such as augmented reality navigation, interconnectedness with the internet and stream media, smartphone integration, high-end assistance systems, or remote services.[23]

Their willingness-to-pay for connected features and additional digital services is primarily correlated with their enthusiasm for novel technologies, superior income levels, and the average amount of driving time. OEMs and providers of marketable connected services should focus on addressing these segments, representing roughly 56% of potential car buyers, by tailoring their portfolio offers according to their preferences and service requirements. The provision of a superior connectivity value proposition is likely to entice customers to switch brands,[24] while above-average purchase prices imply an opportunity to realize higher profit margins.

An in-depth comparison between different types of data services and functionalities revealed that customers are, in general, more eager to switch their manufacturer for driving-related data services and data-based driving functions than for generic data services. Hence, cross-linked navigation, assisted parking, automated collision warning, autopilots, or automatic platooning are higher valued than personalized music and streaming libraries, messaging and news services, or entertainment programs. This cedes a preeminent position to OEMs since driving functions and advanced driving assistance systems are primarily developed and commercialized by automotive manufacturers and their tier-n supplier network.

The economic value of data and its information content has multiplied in today's digital-driven business environment. Companies from every

---
[23] cf. McKinsey & Company (2014) Connected car, automotive value chain unbound S.13–14.
[24] cf. Gutzmer, Alexander (2018) Marken in der Smart City.

Segment	Customer groups with higher interest in and willingness to pay for connectivity features — Preferences and attitudes	Ø willingness to switch for connectivity[1], %	Ø time in car hrs/week	Ø purchase price EUR '000	Segment share, %
Maxed-out car enthusiasts	• Prefer best connectivity features across the board • Clearly want augmented reality navigation • Always want to be connected to the Internet and stream media • Depend on their smartphones a lot • Only very little price sensitive up to EUR 1,800	31	11.3	33	25
Integration and entertainment lovers	• Always want to be connected to the Internet and stream media • Really want their smartphone to integrate into the car • Like advanced navigation features and Google/Apple services • Prefer higher-end assistance sys, e-call, but no health monitoring • Not too price sensitive up to EUR 1,200	27	11.6	26	15
Safe and secure navigators	• Broadly like assistance systems and health monitoring features • Value remote services including vehicle tracking • Prefer internal and car app navigation over external solutions • Not too price sensitive up to EUR 1,200	19	9.2	26	16
Purists/minimalists	• Do not indicate particular interest in any connectivity features • Do not need to be always connected to the Internet • Do not rely on smartphone • Not too price sensitive up to EUR 1,200	11	8.4	22	28
Price-conscious traditionalists	• Do not indicate particular interest in any connectivity features • Tend not to be comfortable with sharing their data • Show high price sensitivity already at EUR 500	7	7.0	18	16
		Ø 19	Ø 9.5	Ø 25	

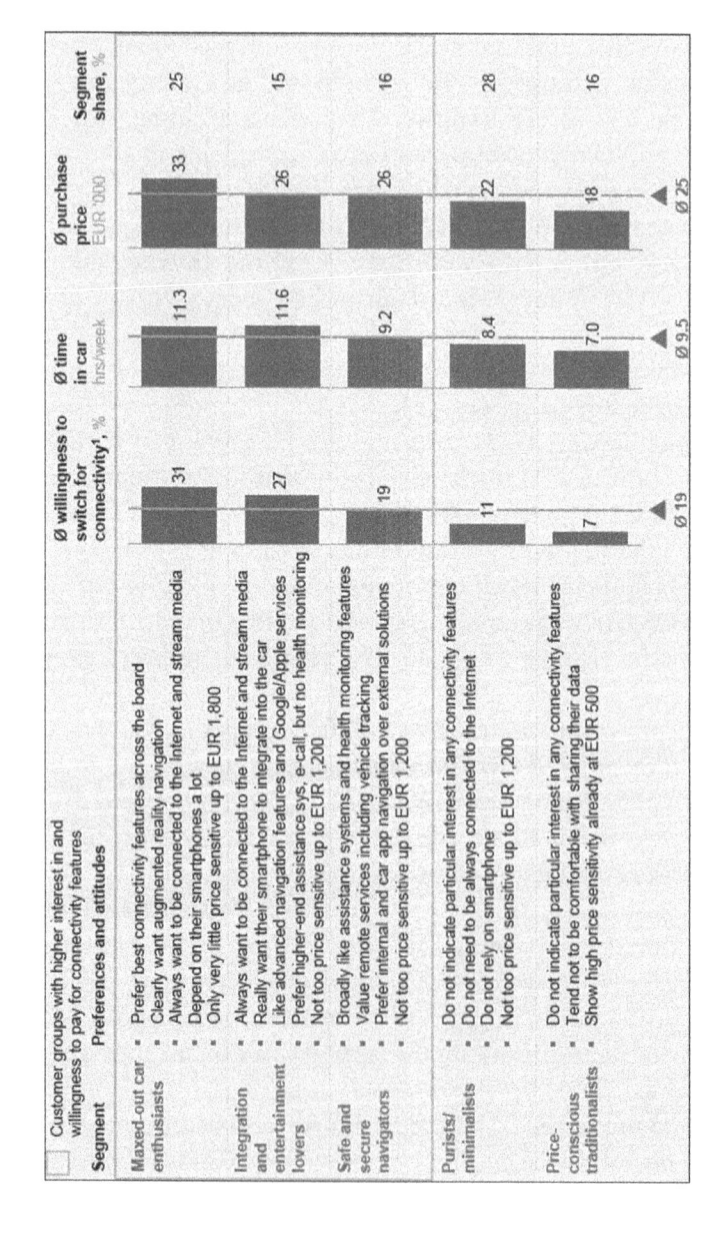

Fig. 4.7 Customer segments that determine the pace of market penetration (McKinsey & Company, 2014, S.12)

industry sector are seeking to gain access to various data streams and databases to generate added value for their product and service portfolios. Beyond use cases of data application in Industry 4.0[25] and various upstream processes in the value-added chain, as indicated in Chap. 3, data access is becoming increasingly crucial in the aftersales area.[26] Usage-based customer data and profiles are the key driver to develop novel digital services and connectivity features. Data as a strategic, intangible asset has been established as a new source of value creation. OEMs and providers of additional in-vehicle services are competing for digital interfaces with their customers to capture and commercialize corresponding data flows.[27] National data privacy policies and General Data Protection Regulations (GDPR), as recently erected by the European Union, impede the unauthorized compilation dissemination of sensitive customer data.[28] Companies seeking the approval of their customers to gather, analyze and use processed data need to ameliorate their public perception and brand image[29] to build customer trust and receive access to data sources.[30] Data scandals, triggered in the telecommunications and social media sector over the past decade, have lasting impacts on the consumers' attitude toward personal data provision and protection. The subsequent surveys about the willingness to share personal and vehicle-related data with OEMs shed light on the potential data collection and acquisition options. In a study published by McKinsey & Company in 2015, 76% of the participants stated they would, subject to certain circumstances, agree to provide data to their automotive manufacturer for product improvement purposes. Roughly 30% of these respondents would insist on a legally binding guarantee that this data is solely used for the aforementioned scope and will not be passed to third parties. Nearly one-quarter would even completely deny data transmission requests.[31]

---

[25] cf. Qusay, Hassan; Khan, Atta; Madani, Sajjad (2018) Internet of Things S.35ff.
[26] cf. McKinsey & Company II (2016) Car data: paving the way to value-creating mobility S.7.
[27] cf. Skilton, Mark (2016) Building Digital Ecosystem Architectures S.51–53.
[28] cf. KPMG II (2017) Global Automotive Executive Survey 2017 S.37–39.
[29] cf. Roland Berger (2011) Automotive landscape 2025 S.41.
[30] cf. Siegfried, Patrick (2020) Marketing- und Vertriebskonzepte von erfolgreichen Unternehmen S. 77ff.
[31] cf. McKinsey & Company III (2015) Wettlauf um den vernetzten Kunden S.19–24.

The studies conducted within the framework of this research work could confirm this conditional reluctance of approving continuative data usage. Based on the same question, 57% of the participants declared they would, in general, rather likely or very likely be willing to provide product-relevant data. Thirty-one percent would refuse any data transmission (Fig. 4.8).

A more specific survey revealed that the vast majority of consumers differentiate between data types based on the perceived privacy sensitivity of the intrinsic information. When asked if they would provide access to data for cost-reduced or gratuitous services such as real-time traffic information or streaming offers, 70% of the respondents stated they would only agree in the case of vehicle-related data.[32] The approval for non-privacy invasive information, including average travel times or frequencies of maintenance would not pose any issue. However, the percentage share of respondents that would even divulge personal data such as location coordinates, driving destinations, or data regarding their driving behavior, was comparably low with a total value of 19.

The illustration below shows different data categories with respective use cases and their corresponding level of perceived privacy sensitivity. It becomes apparent that vehicle-related data such as external road and environmental conditions or vehicle usage information and its technical status are considered less privacy-invasive than personal data[33] and direct HMI communications. Data that is generated through the interaction of the passenger with the vehicle system, however, is crucial to develop certain digital services such as proactive navigation or virtual assistants and to generate revenues through licensing targeted advertisement or e-commerce within the on-board screens.[34] Thus, personal data might become a bottleneck resource if service providers are not able to dispel consumers' doubts about cybersecurity and data protection. Furthermore, customers need to understand what they receive in exchange for their data (Fig. 4.9).

---

[32] cf. McKinsey & Company II (2016) Car data: paving the way to value-creating mobility S.7.

[33] cf. McKinsey & Company; Bloomberg (2016): An integrated perspective on the future of mobility S.24–26.

[34] cf. Institute for Mobility Research (2018) Autonomous Driving - The Impact of Vehicle Automation on Mobility Behaviour S.20.

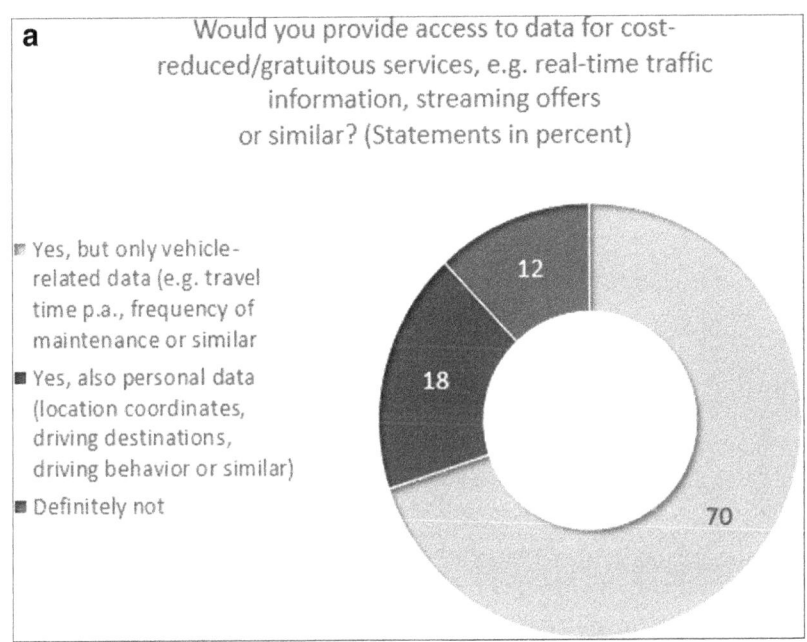

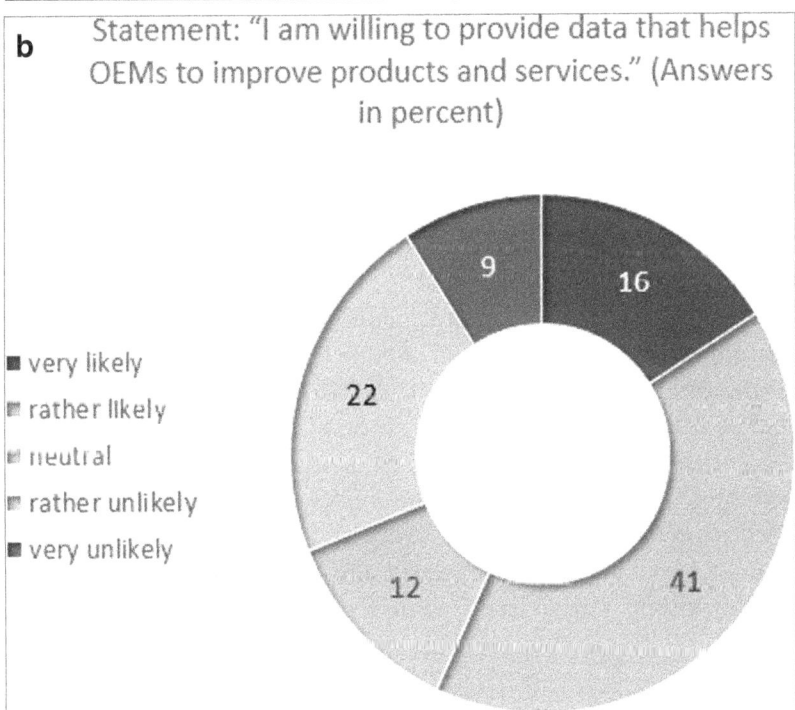

a Would you provide access to data for cost-reduced/gratuitous services, e.g. real-time traffic information, streaming offers or similar? (Statements in percent)

- Yes, but only vehicle-related data (e.g. travel time p.a., frequency of maintenance or similar
- Yes, also personal data (location coordinates, driving destinations, driving behavior or similar)
- Definitely not

12
18
70

b Statement: "I am willing to provide data that helps OEMs to improve products and services." (Answers in percent)

- very likely
- rather likely
- neutral
- rather unlikely
- very unlikely

9
16
22
12
41

Fig. 4.8 Participants' willingness to provide private and vehicle-related data. Source: Author's own representation

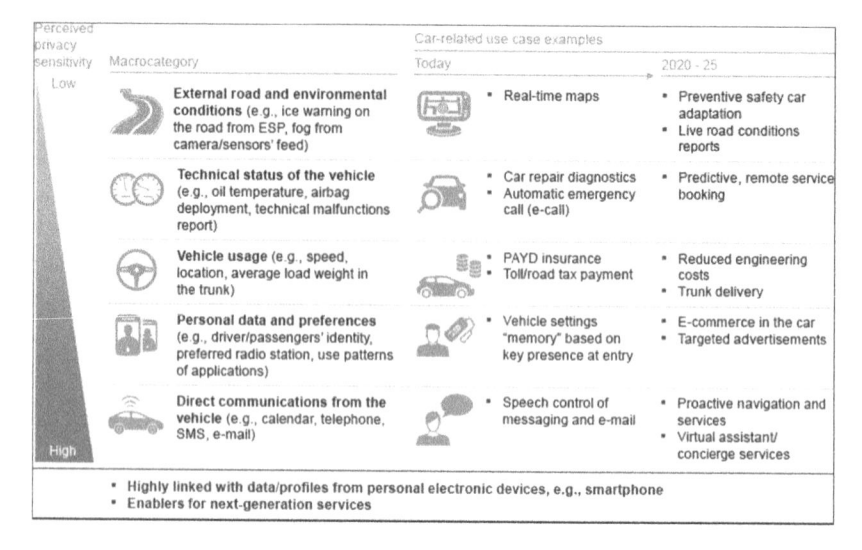

Perceived privacy sensitivity	Macrocategory		Car-related use case examples	
			Today	2020 - 25
Low	External road and environmental conditions (e.g., ice warning on the road from ESP, fog from camera/sensors' feed)		• Real-time maps	• Preventive safety car adaptation • Live road conditions reports
	Technical status of the vehicle (e.g., oil temperature, airbag deployment, technical malfunctions report)		• Car repair diagnostics • Automatic emergency call (e-call)	• Predictive, remote service booking
	Vehicle usage (e.g., speed, location, average load weight in the trunk)		• PAYD insurance • Toll/road tax payment	• Reduced engineering costs • Trunk delivery
	Personal data and preferences (e.g., driver/passengers' identity, preferred radio station, use patterns of applications)		• Vehicle settings "memory" based on key presence at entry	• E-commerce in the car • Targeted advertisements
High	Direct communications from the vehicle (e.g., calendar, telephone, SMS, e-mail)		• Speech control of messaging and e-mail	• Proactive navigation and services • Virtual assistant/concierge services

• Highly linked with data/profiles from personal electronic devices, e.g., smartphone
• Enablers for next-generation services

Fig. 4.9 Macro-categories of data with different levels of perceived privacy sensitivity (McKinsey & Company II, 2016, S.16)

However, OEMs and service providers need to differentiate the consumers' interest in data privacy between the key sales markets. The cultural comprehension of privacy and personal rights represents a determining factor in gaining access to sensitive data. According to a study conducted by McKinsey & Company, Chinese respondents were more likely to transmit both vehicle- and individual-related data.[35] Therefore, service providers may choose to initially introduce new services and features on the Chinese market before approaching a comprehensive commercialization on a global level. In consequence, China could become the incubator of the car data "revolution."[36]

As the previous sections have argued, automotive manufacturers are confronted with an upheaval of the industry due to customers' altered mobility requirements and their emerging faible for connected features. Beyond new challenges within the framework of data handling, commencing cannibalization based on novel mobility solutions, and the value

---

[35] cf. McKinsey & Company II (2016) Car data: paving the way to value-creating mobility S.7–9.
[36] cf. PwC - Strategy& (2016) Connected car report 2016 S.20ff.

that consumers attach to innovative connectivity, OEMs are facing a decline of the customer group possessing the, on average, highest purchasing power.[37] The baby-boomer generation that comprises people born between 1965 and 1980 will reach interim pension or retirement until 2020, thus shedding their economic importance as solvent buyers and causing a regression of the potential customer base by 1.5 million. A progressing urbanization and the corresponding demand for product specifications adapted to the individual lifestyle habits will moreover lead to an increasing demand for smaller cars.[38]

In the battle for customers, OEM still enjoys a higher reputation. Two surveys conducted by the Boston Consulting Group[39] and Deloitte[40] in 2016 respectively 2017 revealed, that on average, 46% of the respondents prefer established automotive manufacturers as producers of self-driving vehicles, followed by tech companies, new car manufacturers, government-owned and supervised companies, and automotive suppliers with percental shares of 16% to 5%. Aside from the sole hardware development perspective, customers have more confidence in OEMs to assure a responsible handling of data security and protection issues compared to renowned software providers such as Apple, Google or Microsoft. While automotive producers hold a leap of faith in the German market, Chinese respondents clearly indicated their trust towards technology and software ventures, coercing OEMs to collaborate with the aforementioned companies and start-ups to gain access to an extended customer base (Fig. 4.10).

According to the survey results that have been analyzed within the context of this research work, the respondents rated US-based technology companies and German automotive manufacturers, headed by Mercedes-Benz, as most competent to develop autonomous vehicles, while primarily Asian OEMs bring up the rear.[41] Arguments that support this appraisal are mostly based on the accumulated knowledge and core competencies

---

[37] cf. PwC; NTT Data Deutschland (2013) Automotive Retail S.21–23.
[38] cf. PwC (2018) Five trends transforming the Automotive Industry S.9–12.
[39] cf. The Boston Consulting Group; World Economic Forum (2016) Self-driving vehicles, robot-taxis, and the urban mobility revolution S.7–11.
[40] cf. Deloitte II (2017) The race to autonomous driving S.7–11.
[41] cf. L F K. (2017) Mapping the Road to Autonomous Vehicles S.5.

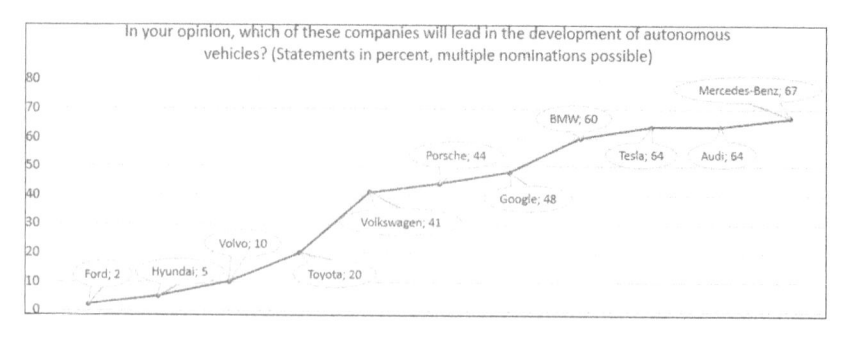

Fig. 4.10 In your opinion, which of these companies will lead in the development of autonomous vehicles? Source: Author's own representation

of established automotive producers as well as on the financial and innovative capacities of tech giants.[42]

However, OEMs further need to defend themselves against rivals that are entering the market for mobility-as-a-service solutions luring potential customers away from private vehicle ownership. Indeed, several German automotive manufacturers have already introduced own car sharing models or multi-modal transportation platforms such as car2go (Mercedes-Benz), DriveNow (BMW), or Moovel (Volkswagen)[43] which aim at generating additional revenues beyond basic vehicle sales but are attended on a risk of cannibalizing the traditional business model profitability. The aforementioned survey shows that on-demand and shared mobility models, in fact, do not represent a significant market share in the overall mobility sector with 20% of—more or less—frequent users, however, considering these services have only recently been introduced to the market, their growth rate and market diffusion are progressing inexorably (Fig. 4.11).

The study published by Deloitte in 2017 moreover indicated that approximately 45% of consumers in emerging countries with large metropolitan areas and low shares of private car ownership such as India and China, are using car sharing services at least once a week.[44] However, the

---

[42] cf. KPMG II (2017) Global Automotive Executive Survey 2017 S.18.
[43] cf. McKinsey & Company II (2015) Urban mobility at a tipping point S.15.
[44] cf. Deloitte III (2017) What's ahead for fully autonomous driving S.4–10.

4 Urban Mobility Revolution: A Quantitative Analysis

How often do you use the following means of transport?
(Statements in percent)

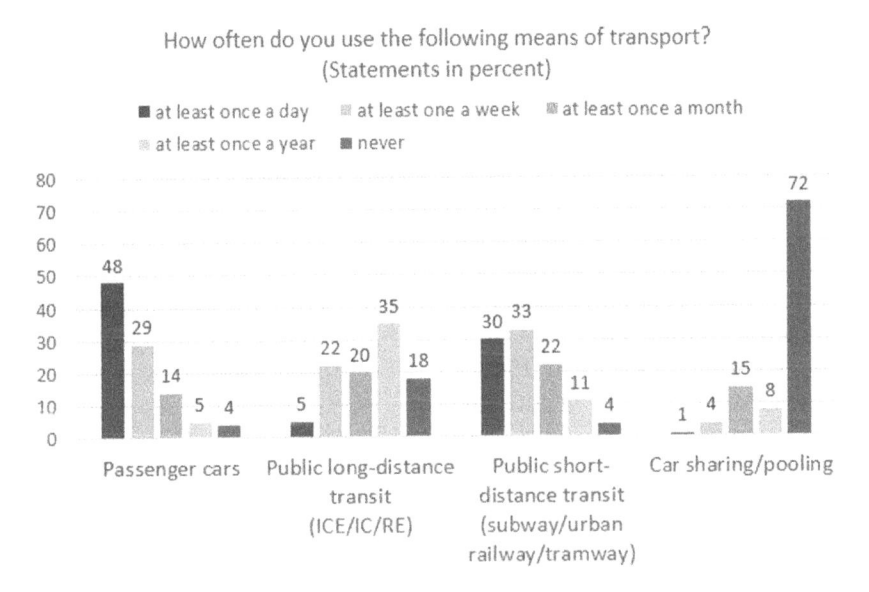

- at least once a day
- at least one a week
- at least once a month
- at least once a year
- never

Means of transport	at least once a day	at least one a week	at least once a month	at least once a year	never
Passenger cars	48	29	14	5	4
Public long-distance transit (ICE/IC/RE)	5	22	20	35	18
Public short-distance transit (subway/urban railway/tramway)	30	33	22	11	4
Car sharing/pooling	1	4	8	15	72

Fig. 4.11 How often do you use the following means of transport? Source: Author's own representation

users of these mobility options hardly jeopardize the traditional business model since private vehicle purchase typically exceeds the financial capacity of those customer segments but represent an opportunity to extend the commercialization of company-owned mobility services.

In conclusion, value propositions of newly released vehicles must be more customer- and demand-oriented to realize a promising positioning within the competitive environment. New-car purchases increasingly relocate towards compact and small-premium segments. Since private vehicle ownership incrementally loses its sole value as a stimulation of mobility needs but moreover serves as means for self-expression, a need for brand differentiation is in the offing.[45] As novel innovations and technology features are no longer exclusively introduced through a top-down approach, the boundaries between premium and volume segments are becoming indistinct. Mass market producers should thus emphasize their brand experience and differentiate their product portfolio towards more

---
[45] cf. Diehlmann, Jens; Häcker, Dr. Joachim (2013) Automotive Management S.10.

lifestyle-oriented vehicles which have already captured considerable market shares throughout the previous years.

4.2 Value of Time and Willingness-to-Pay for Digital Services

Mobility, as it is known today, is at the crossroads. Individual mobility is facing a new stage of development that will have a game-changing impact on the entire industry and the form of locomotion. The progressive commercial launches of gradually more automated vehicle functions enable the driver and his passengers to use the newly reclaimed time of travel for alternative activities in the future. The creation of a so-called "third space" between home and work brings the opportunity to introduce accompanying services within the vehicle.[46] The extent of disposable time or respectively the degree of complexity that digital services and activities require is constrained by the different levels of automation that have been defined by the National Highway Traffic Safety Administration (NHTSA) and the Association of the Automotive Industry (VDA) (Fig. 4.12).[47]

The scale is grouped into monitored and non-monitored driving levels. Vehicles on levels 0–2 require the driver to monitor the systems at all times since he resumes the entire liability of any traffic and driving incidents inflicted by his own driving behavior.[48] While level 0 vehicles without any form of automation were the common standard for over one century, assisted and partially automated vehicles on levels 1 and 2 dispose of function-specific automation features. The driver is supported by longitudinal or lateral control functions such as lane assist, blind-spot detection, adaptive cruise control, or pedestrian detection.

In a level 2 vehicle these specific driving functions can be operated simultaneously, that is to say, at least two steering or detection functions are transferred concurrently to the vehicle's control system.[49] Conditionally

---

[46] cf. Humboldt-Universität zu Berlin (2016) User Perspectives on Autonomous Driving S.30ff.
[47] cf. National Highway Traffic Safety Administration (N.A.) Automated Vehicles for Safety.
[48] cf. L.E.K. (2017) Mapping the Road to Autonomous Vehicles S.1.
[49] cf. OECD/International Transport Forum (2015) Automated and Autonomous Driving S.14.

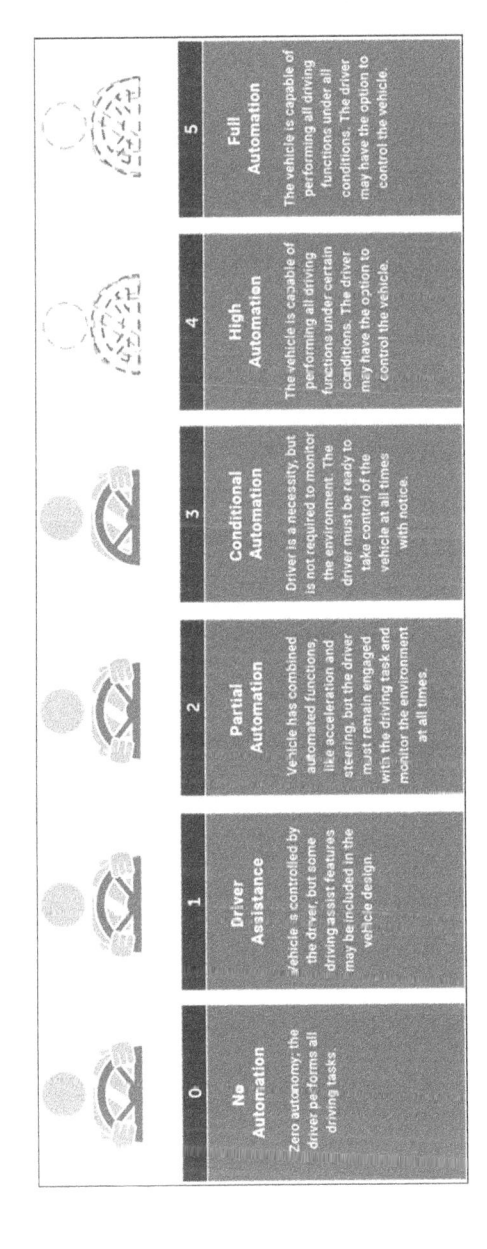

4 Urban Mobility Revolution: A Quantitative Analysis 75

0	1	2	3	4	5
No Automation	Driver Assistance	Partial Automation	Conditional Automation	High Automation	Full Automation
Zero autonomy; the driver performs all driving tasks.	Vehicle is controlled by the driver, but some driving assist features may be included in the vehicle design.	Vehicle has combined automated functions, like acceleration and steering, but the driver must remain engaged with the driving task and monitor the environment at all times.	Driver is a necessity, but is not required to monitor the environment. The driver must be ready to take control of the vehicle at all times with notice.	The vehicle is capable of performing all driving functions under certain conditions. The driver may have the option to control the vehicle.	The vehicle is capable of performing all driving functions under all conditions. The driver may have the option to control the vehicle.

Fig. 4.1 Society of Automotive Engineers (SAE) Vehicle Automation Levels. National Highway Traffic Safety Administration (NHTSA) (n.a.)

automated vehicles on level 3 offer certain autonomous driving modes that enable the car to conduct all safety-relevant functions in the automobile steering. Hence, features such as highway chauffeur, active parking assistant systems, or emergency programs do not require permanent system monitoring. According to the NHTSA, level 4 vehicles will reach market maturity between 2020 and 2030. In this scenario, all vehicle functions are operated autonomously and all safety-relevant driving situations are conducted self-dependently by the vehicle system. The last stage of expansion in technological development is, according to the VDA, driverless cars that do not require an attended driver since the system can cope with all situations automatically during the entire journey.[50] Today's stages of development comprise vehicles of levels 0–3.

In a medium-term vision, the progressing automation enables ancillary activities whose complexities are dependent on the degree of freedom provided by the vehicle-integrated autonomous technology.[51] These activities can be ascribed to the fulfillment of underlying needs. The goal of the subsequent assessment and the intention of OEMs and third parties involved in the automotive business landscape is the identification and (financial) evaluation of specific activities that are conducted by the driver during the journey time in autonomous or highly automated vehicles.[52] To enable a valuation of (digital) service potentials, possible ancillary activities have been categorized within in the context of this study. Based on six fundamental needs, service groups were deducted which, in turn, comprise concrete activities. The study "Value of Time" published by Horváth & Partners[53] which served as concept approach for the setup of the quantitative study conducted for this research work further measured the relevance and frequency of use of certain service clusters. Subsequent to the potential usage survey, participants were asked to indicate their service-specific monthly willingness-to-pay and to allocate this value on the underlying service groups on a percentage base. In the end, service-specific payment preferences, the willingness to accept surcharges

---

[50] PA Consulting (2017) Autonomous Driving S.5.
[51] cf. A.T: Kearney (2016) S.10.
[52] cf. Bundesverband Digitale Wirtschaft (2016) Connected Cars – Geschäftsmodelle S.8–10.
[53] cf. Horváth & Partners; Fraunhofer (2016) The Value of Time S.20–28.

4 Urban Mobility Revolution: A Quantitative Analysis

for the basic provision of the services, and the individual values of time were inquired.

The key figure "Value of Time" specifies how much one additional hour of disposable time is worth to each person in a monetary value. Thus, the VoT is an immaterial, individual, and situational asset with price information that is highly dependable on personal preferences, driving purpose, number of passengers, or the particular situation.

The assessment has revealed that the VoT accumulated over the entire vehicle life cycle predominantly equiponderates the costs for the integration of the autonomous driving function and the concomitant access to additional digital services.[54] The provision of associated services, therefore, holds a significant commercial yield. The highest value can be ascribed to young respondents and higher-income earners. In conclusion, the autonomous driving functions bear a high strategic relevance for OEMs competing for customers' favor (Fig. 4.13).

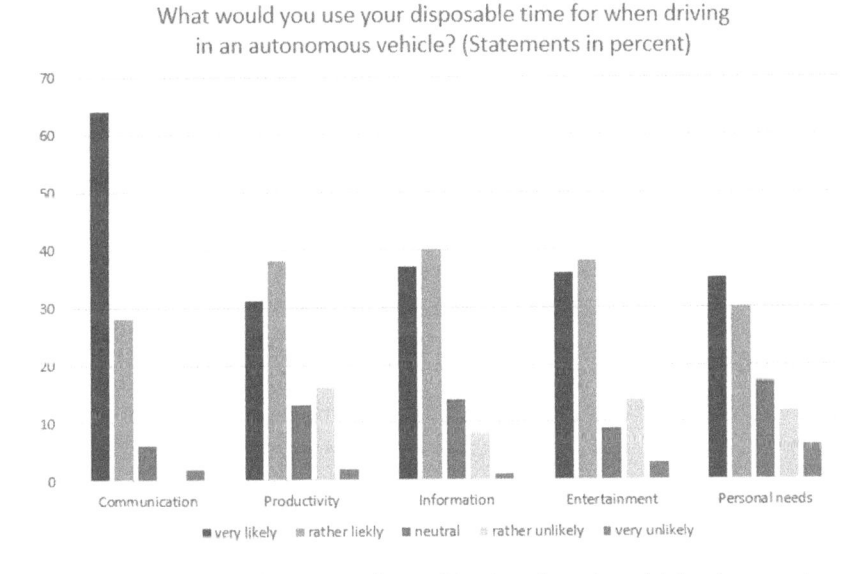

Fig. 4.13 What would use your disposable time for when driving in an autonomous vehicle? Source: Author's own representation

---

[54] cf. Accenture (2015) Autonomous driving - are OEMs losing the driver seat? S.5–7.

The subsequent section summarizes the key findings of the studies. The majority of users, roughly 75% across all specified user groups, are willing to pay for value-added services. A more differentiated analysis of the individual willingness-to-pay according to respondents' age groups, social and geographical background shows significant disparities in the tolerated amount of extra charges depending on demographic influencing factors.[55]

Based on the evaluation of the relevance and willingness-to-pay for the activities associated to the superordinate service groups, the categories "Communication," "Productivity," and "Information"[56] were ranked the most important. Ninety-two percent of the survey participants stated that they would very like or rather likely use services that are related to communication purposes such as e-mail features, phone integration, social media applications, or virtual chat rooms. Productivity services that include various occupational, online banking, or educational applications would attract interest of 69% of the respondents. Information-related services such as traffic routing, virtual product testing, or news updates would potentially be acquired by 77% of autonomous vehicle users.

Based on this result, it can be inferred that users want to relocate obligatory and, in general, value-adding activities into the vehicle. For these activities, users agree to pay more compared to non-obligatory, entertaining services. Value-added services need to be differentiated by the degree of vehicle function automation and general willingness-to-pay. This offers the opportunity to identify focal points in the future development of service portfolios. Moreover, in combination with the inquired relevance of service groups, requirements and specifications of autonomous vehicles can be deducted irrespective of the particular willingness-to-pay. On the basis of the survey results, a significant market potential can be derived which is not only decisive for automotive manufacturers and tier-n suppliers but also for emerging start-ups and incumbent IT companies seeking to receive access to new vehicle-related revenue streams.[57] These

---

[55] cf. Gerdes, Christian; Maurer, Markus (2015) Autonomous Driving S.9.
[56] cf. Horváth & Partners; Fraunhofer (2016) The Value of Time S.20ff.
[57] cf. Bain & Company (2017) An Autonomous Car Roadmap for Suppliers.

players need to differentiate dissimilar customer clusters to tailor their service offerings accordingly since the willingness-to-pay is correlated with various factors such as the average daily driving time. The willingness-to-pay proceeds disproportionately with the duration of vehicle usage, that is to say, the initial willingness to accept surcharges is high whereas the curve flattens with increasing price levels, similar to the slope of a saturation curve.[58] Another conclusion is that the usage frequency of service offerings is determined by the customers' affiliation to different generations. The increasing usage of full and hybrid service offerings has developed as a continuous trend in the past 10–20 years.[59] While physical products and intangible services have formerly been separate goods, there has been an incremental blurring of boundaries between the kinds of goods in the last decades. Younger users who have grown up with this form of product comprehension are accustomed to situationally pay for additional services and features ensuing the actual purchase of the product.

Well-known examples for Pay-as-you-Drive (PayD) or flat-rate-based service models are on-demand streaming services such as Spotify, Netflix, or Amazon Prime Video. Coming back to the tangible hardware of the vehicle, it can be stated that the willingness-to-pay for optional equipment is primarily dependent on the respective car segment. Typically, buyers of upscale vehicles such as Sports Utility Vehicles (SUV) or sports cars are prone to accept higher surcharges compared to sales leads of smaller car classes.[60] These purchase patterns are mainly retraceable to the average income of the buyers of the particular segments.

But in principle, it can be stated that the pricing of the service portfolio applicable in autonomous vehicles should be configured independently from the vehicle segment. If services are provided with the same price conditions to all customers, these services can be developed across-the-board thus reaching a large consumer pool in the early stages. A well-tried procedure of the automotive industry has been a top-down approach

---

[58] cf. IBM Institute for Business Value (2016) A new relationship – people and cars S.15ff.
[59] cf. Hülsmann, Michael; Fornahl, Dirk (2014) Evolutionary Paths towards the Mobility Patterns of the Future S.38–40.
[60] cf. Horváth & Partners; Fraunhofer (2016) The Value of Time S.24–26.

to market innovations exclusively through classic carriers of technology in the luxury and upper classes with a delayed diffusion of these features in smaller car segments. The indicated inclination of the willingness-to-pay extra for the autonomous driving function, though, enables OEMs to introduce the technology directly within the voluminous compact and medium-sized vehicles classes thus quickly generating scale effects and a competitive advantage.[61] However, the low amount that respondents were willing to pay extra compared to the high costs of production suggests that automotive manufacturers should pursue a low-price penetration strategy to break the buying resistance and increase revenues through scale effects and an enlarged service portfolio.[62]

4.3 Valuation of the Future Market Volume

In order to forecast the future automotive market volume and different revenue pool sizes on an accurate level, it is conducive to recapitulate the most relevant influencing factors, both negative and positive. As the previously indicated study key findings have shown, connectivity features and autonomous vehicles attract interest in the populace, though demographic factors such as generation affiliation, gender, income levels, or place of residence need to be accounted for. In particular, young customers, high-income earners, and inhabitants of urbanized regions represent the most promising consumer segments, especially when considering their function as early adopters that contribute to an increasing market penetration and the diffusion of novel digital services and features.

Additional revenue pools can be captured by growing adoption rates of electrified and hybrid or plug-in vehicles[63] as well as by the provision of alternative mobility solutions such as car sharing or multi-modal transportation platforms[64] introduced by several OEMs and third parties. Growing middle classes in emerging countries, predominantly in BRIC

---

[61] cf. KPMG I (2018) Autonomous Vehicle Readiness Index S.24.
[62] cf. McKinsey & Company III (2016) Monetizing car data S.25–28.
[63] cf. Diez, Willi (2018) Wohin steuert die deutsche Automobilindustrie? S.104–106.
[64] cf. Meyer, Gereon; Shaheen, Susan (2017) Disrupting Mobility S.149.

states, constitute new profitable outlet markets that compensate for stalling sales volume in Western and industrialized countries. But there is also the other side of the coin. As indicated in Chap. 2, the attitude towards private car ownership and private mobility is incrementally altering due to TCO considerations, environmental awareness, and the upcoming consumer desire for flexible, cost-efficient urban locomotion. The continuing relocation of innovative vehicle features and connected components into smaller car segments will successively shift the sales structure towards low-margin revenue pools. In addition, the overall profit margin of the majority of OEMs and tier-n suppliers has already suffered from increasing competition in the globalized business and the customers' demand for individual, personalized products reinforcing the complexity of the value chain and worldwide logistics network.

In a long-term perspective, the future market volume is determined by four business areas, to be specific, new-car sales, the economic significance of e-mobility, mobility-as-a-service[65] business models, and digital services provided in the "third space," entailing nonrecurring, subscription- or flat-rate-based fees.[66] The subsequent diagrams show the forecasted, cumulative sales volumes for autonomous, electrified and connected vehicles in the United States, the European Union and China for the previous year as well as for the years 2020, 2025, and 2030 (Fig. 4.14).

The first bar chart is sub-divided by the different levels of vehicle automation determined by the NHTSA and VDA.[67] It becomes apparent that vehicles with higher levels of autonomy will progressively penetrate the markets while level 0–2 car models will eventually vanish from the automotive landscape. The octuplication of highly and fully autonomous vehicles between 2020 and 2030 is primarily induced by advances in technological development leading to the marketability of AVs and thus, in turn, triggering the emergence of robo-taxi models and respective autonomous vehicle fleets.[68]

---

[65] cf. Winkelhake, Uwe (2017) Die digitale Transformation der Automobilindustrie S.299–300.
[66] cf. Seipp, Vanessa; Michel, Alex; Siegfried, Patrick (2020) Review of International Supply Chain Risk Within Banking Regulations in Asia, US and EU Including Proposals to Improve Cost Efficiency by Meeting Regulatory Compliance S.25ff.
[67] cf. National Highway Traffic Safety Administration (N.A.) Automated Vehicles for Safety.
[68] cf. PwC - Strategy& (2017) The 2017 Strategy& Digital Auto Report S.14.

Fig. 4.14 Forecast of new car sales in the EU/US and China (PwC's Strategy&, 2017, S.8)

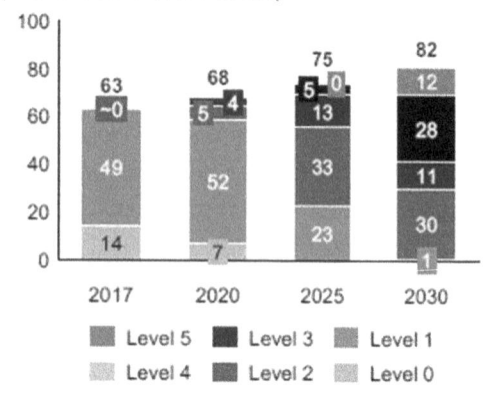

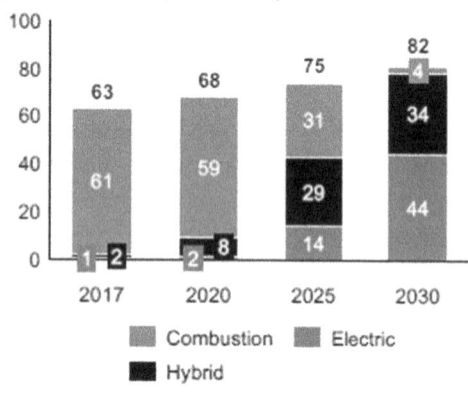

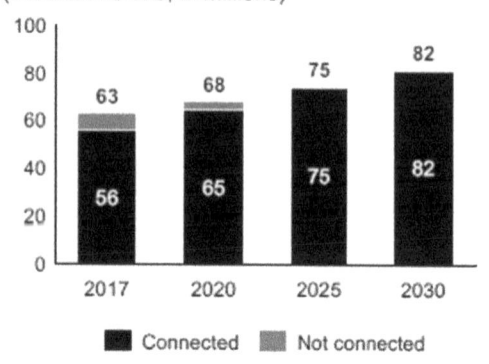

4 Urban Mobility Revolution: A Quantitative Analysis

The second diagram shows the distribution of new vehicles sales based on their integrated powertrain in the three main business markets up to the year 2030. Vehicles powered by combustion engine face a dismal foreboding with a negative Compound Annual Growth Rate (CAGR) of 18.9%. Electric and hybrid vehicles, however, will chronicle exceedingly positive CAGRs of 33.79 and 24.35% respectively. According to this forecast, roughly 95% of vehicles sold in 2030 will be affiliated to alternative powertrains.[69] This upheaval in the automotive landscape will be driven by a strong legislative push that has already been initiated by the majority of emerging and industrialized countries and will progressively be expediated, potentially even climaxing in prohibitions for combustion engines from 2030 on. Many governments have announced associated policies and threshold values for the allocation of registration shares regarding fuel-powered and electrified vehicles (Fig. 4.15).[70]

Further decisive reasons for the ascent of environmentally sustainable powertrains are that components for electric power systems, in particular

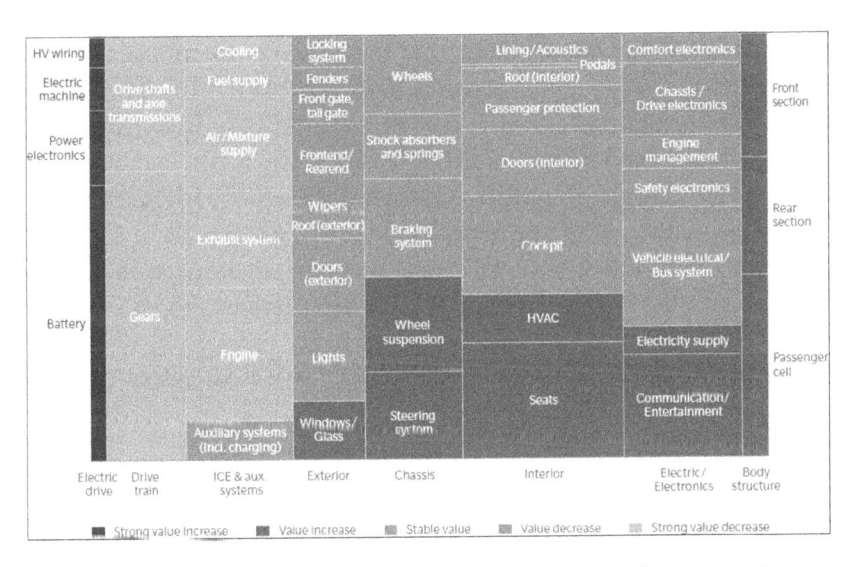

Fig. 4.15 Total market value split of vehicle modules and future development (Oliver Wyman, 2017,␣33)

---

[69] cf. PwC - Strategy& (2017) The 2017 Strategy& Digital Auto Report S.18–20.
[70] cf. Roland Berger (2011) Automotive landscape 2025 S.28–30.

lithium-ion batteries, will reach a price tipping point around 2025[71] while the development of the required charging infrastructure will meet a sufficient level. In consequence, initial and running costs as well as former range limitations will no longer impede the competitiveness of alternative powertrains.

The figure above shows the total market value split of vehicle modules and accentuates the future development through the blue shading. E-mobility and its associated components such as high-voltage wiring, electric machines, power electronics, and batteries, are ascribed a significant future value growth while combustion-engine-related components like exhaust systems, fuel supply, drive shafts or axle transmissions that are currently capturing a high local value,[72] will pale in comparison. This qualitative appraisal of the future relevance of vehicles modules further underpins the forecasted market penetration of alternative powertrains.

The bar chart on the right (on the previous page), illustrating the diffusion of connectivity features in new-sold cars, explicitly indicates the extinction of vehicles that are not cross-linked with the Internet of Things. Emanating from roughly 9% of non-connected vehicles in the previous year, this number is likely to decline to 0% between 2020 and 2025.[73] The legal and customer pull for connectivity will lead to an automotive landscape entirely and homogenously composed of connected vehicles (Fig. 4.16).

The table above summarizes the forecasted installed vehicle base for each sales market until the year 2030. Based on this data, four main theses can be erected, that will have a significant impact on the sales structure of OEMs and, consequently, tier-n suppliers (Fig. 4.17):

	2017			2020			2025			2030		
	Autonomous	Electric	Connected	Autonomous	Electric	Connected	Autonomous	Electric	Connected	Autonomous	Electric	Connected
U.S.	-	0.5	31.3	-	2.2	67.3	2.1	11.3	116.3	20.8	45.0	146.0
E.U.	-	0.8	32.6	-	1.5	71.3	2.7	9.5	123.5	27.1	45.4	147.7
China	-	1.2	27.8	-	4.0	99.2	2.4	20.5	230.9	33.1	73.7	299.0
Total	-	2.5	91.7	-	7.6	237.7	7.3	41.2	470.7	81.0	164.0	592.7

Fig. 4.16 Installed vehicle bases from 2017 to 2030 (in millions) (PwC's Strategy&, 2017, S.9)

---

[71] cf. Hülsmann, Michael; Fornahl, Dirk (2014) Evolutionary Paths towards the Mobility Patterns of the Future S.38–40.
[72] cf. Oliver Wyman (2017) The Oliver Wyman Automotive Manager S.33.
[73] cf. Chen, Yaobin; Li, Lingxi (2014) Advances in Intelligent Vehicles S.60–68.

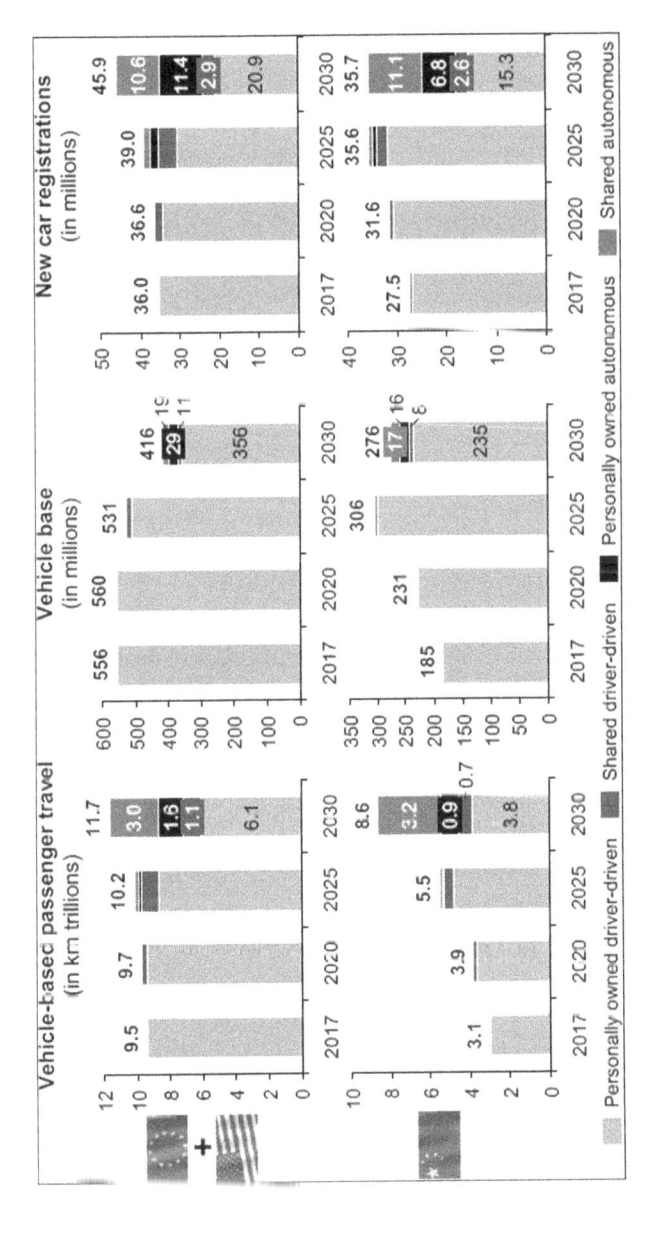

4 Urban Mobility Revolution: A Quantitative Analysis 85

Fig. 4.17 Forecasted mobility development in the EU/US and China (PwC's Strategy&, 2017, S.18)

1. There will be a progressive move to shared and autonomous mobility driven by policy and technology breakthroughs.
2. The overall average distance driven and the relative share of vehicle-based mobility will grow, especially in China.
3. Vehicle capacity utilization and turnover will increase due to emerging car sharing and pooling business models.[74]
4. In turn, shared mobility solutions will lead to a decline in the total vehicle base in a long-term perspective, while temporarily increasing overall new car sales.

This diagram compilation illustrates the theses stated above. The number of vehicle-based driven kilometers per year will rise by 23.16% until 2030 due to increasing mobility needs, an enlarged car user base and shared mobility services. Experts assume that the amount of distances driven in China will surpass individual levels of the European Union and the USA around 2025.[75] Secondly, the overall vehicle base in the EU and USA will incrementally decline, emanating from a higher utilization rate of shared and autonomous vehicles, resulting in a negative compound annual growth rate of roughly 2.2%. In China, however, this trend will set in time-delayed since private car ownership still holds high significance value and personally owned autonomous vehicles represent a prestige good.[76] But in general, the strong increase in shared mobility may lead to a drop in vehicles sales during the transformation before higher turnover rates push the sales figures.

In order to gain a deeper insight into the future development of the third business area, the mobility-as-a-service models, the bar charts below show the temporal growth of market size for each distinct sales market (Fig. 4.18).

The key underlying driver for the ascent of mobility services is the increasing global vehicle-based passenger travel that results from a significant price erosion[77] for personal mobility options. Improved efficiency

---

[74] cf. Goodyear; London School of Economics II (2015) Millennial's views on the future of mobility in Europe S.12.
[75] cf. Roland Berger; fka Forschungsgesellschaft (2016) Index "Automatisierte Fahrzeuge" S.18–30.
[76] cf. PwC - Strategy& (2016) Connected car report 2016 S.17–20.
[77] cf. PwC (2018) Five trends transforming the Automotive Industry S.33–36.

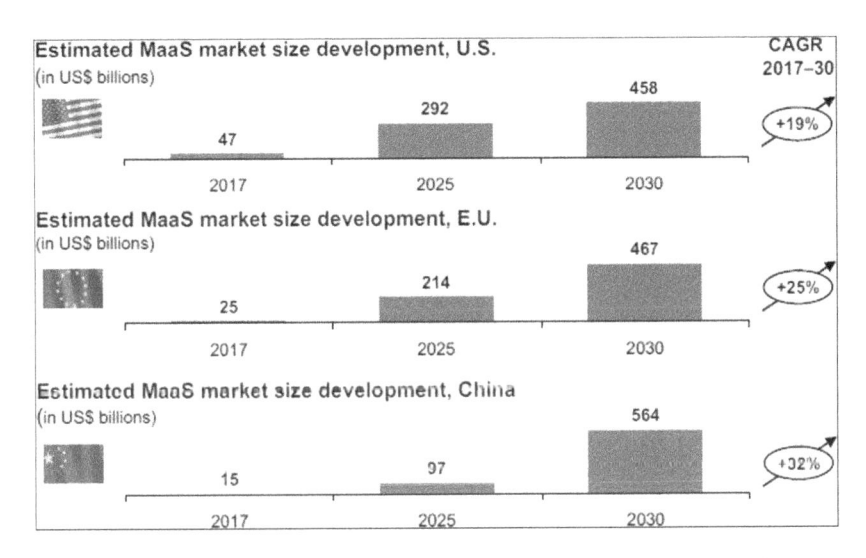

4 Urban Mobility Revolution: A Quantitative Analysis

Estimated MaaS market size development, U.S. (in US$ billions)
- 2017: 47
- 2025: 292
- 2030: 458
- CAGR 2017–30: +19%

Estimated MaaS market size development, E.U. (in US$ billions)
- 2017: 25
- 2025: 214
- 2030: 467
- CAGR 2017–30: +25%

Estimated MaaS market size development, China (in US$ billions)
- 2017: 15
- 2025: 97
- 2030: 564
- CAGR 2017–30: +32%

Fig. 4.18 Estimated Mobility-as-a-Service market size developments in the EU/US and China (PwC's Strategy&, 2017, S.19)

and maintenance of autonomous vehicles as well as the intensification of car sharing and pooling services effectuate a decline of the average price level for one driven distance unit.[78] In conclusion, shared mobility will reach a total value of roughly 1500 billion US Dollars in the three largest consumer markets with a combined growth rate of 24% p.a.

This price development becomes even more apparent when throwing a glance at the distribution of Western premium household expenditures on mobility options, both in a 2030 scenario and a pure-play shared scenario (Fig. 4.19).

The 2030 scenario, assuming that 35% of annual kilometers driven are attributed to shared mobility services, discloses stagnating industry value creation respectively spending related to OEMs and suppliers whereas the expense groups gas, used vehicles, public transport, and MaaS show negative compound annual rates of 1.32%, eventually resulting in a total regression of 15.90% until 2030. In this forecast, the profit generated by automotive manufacturers and suppliers will drop by 12.90% while the overall profit pool decreases by approximately 7.8%. A look at the

---
[78] cf. Helmond, Marc; Terry, Brian (2016) Lieferantenmanagement 2030 S.40ff.

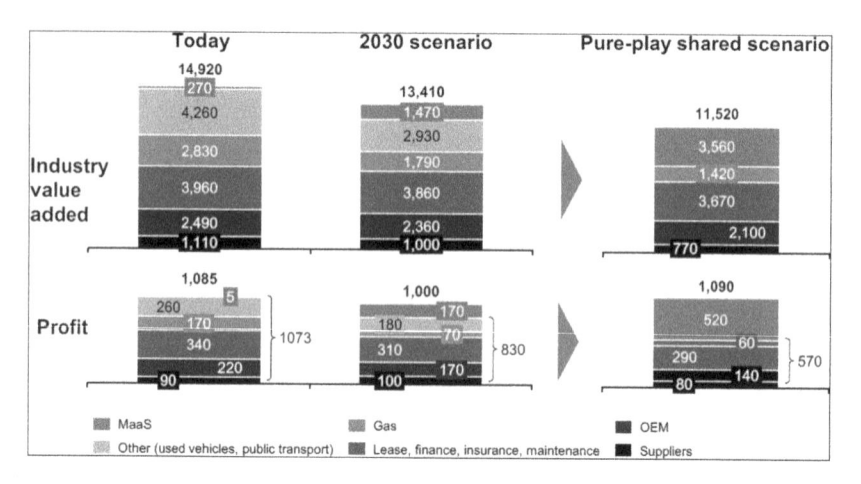

Fig. 4.19 Industry distribution of household mobility expenditure in shared autonomous scenarios, Distribution of Western premium household spend (in US$) (PwC's Strategy&, 2017, S.26)

hypothetical pure-play shared scenario even worsens the prospects of traditional OEMs and tier-n suppliers. The industrial value creation and the cumulative profit pools are predicted to diminish by 20.3%, respectively 29%. These numbers illustrate the initial paradox relation between value creation and profit margin regarding traditional automotive manufacturing and mobility-as-a-service models.[79] Although the average household spending on MaaS options will decline due to assumed price erosions, the economic rate of return will rise based on scale effects and a steep learning curve.[80] Lower household expenditures on OEM- and supplier-related products, however, will negatively affect the overall profitability which cannot even be compensated for by reduced development or production costs, not accounting for potential in-car featured digital services offered by automotive players that would yield further revenues. In conclusion, OEMs should further invest in the provision of company-owned MaaS solutions and advanced, vehicle-related digital services to

---

[79] cf. McKinsey & Company II (2016) Car data: paving the way to value-creating mobility S.12.
[80] cf. McKinsey & Company (2014) Connected car, automotive value chain unbound S.20–23.

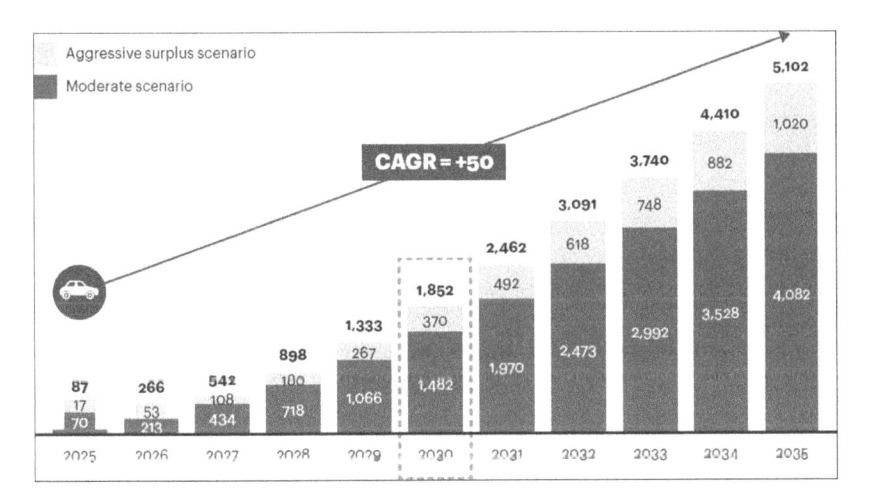

Fig. 4.20 Global self-driving minutes (long-distance commuting only) (A.T. Kearney, 2016, S.9)

countervail the impairing trend of declining traditional value creation and shift of profitability structures.

As already indicated in the previous sub-chapter, autonomous and highly automated vehicles will capacitate their passengers to pursue activities not related to the driving task.[81] The subsequent appraisal reckons with 1482 trillion minutes of global idle time in 2030, translating to roughly 28,200 centuries of disposable time in just 1 year (Fig. 4.20).

The creation of this so-called "third space" represents a huge opportunity to commercialize supplementary digital services that generate continuous revenues from subscription-, flat-rate-based or nonrecurring fees.[82] The market volume of these additional services will observe steep economic growth rates that are positively correlated with the amount of global idle time.

---

[81] cf. Karls, Ingolf; Mueck, Markus (2018) Networking Vehicles to Everything S.112ff.
[82] cf. Kosch, Timo et al. (N.A.) Automotive Internetworking.

References

Accenture. (2015). *Autonomous driving—are OEMs losing the driver seat?* S. 1–8. Retrieved March 9, 2021, from, https://www.accenture.com/t20170517 T041153__w__/cn-en/_acnmedia/PDF-51/Accenture-PoV-Autonomous-Driving.pdf

Anderson, J., Kalra, N., Stanley, K., Sorenson, P., Samaras, C., & Oluwatola, O. (2016). *Autonomous vehicle technology. A guide for policymakers.* Rand Corporation.

Bain & Company. (2017). *An autonomous car roadmap for suppliers*, S. 1–8. Retrieved March 9, 2021, from, http://www.bain.de/Images/BAIN_BRIEF_An_Autonomous_Car_Roadmap_for_Suppliers.pdf

BearingPoint II. (2017). *Mensch & Maschine im Kundenservice: Traumpaar statt Konkurrenz! Automatisiert, proaktiv und persönlich: Der richtige Mix für den Kundenservice der Zukunft*, S. 1–23. Retrieved May 11, 2021, from, https://www.bearingpoint.com/de-de/unsere-expertise/insights/kundenservice/

Blanke, T. (2014). *Digital asset ecosystems. Rethinking crowds and clouds.* Chandos Publishing.

Bundesverband Digitale Wirtschaft. (2016). *Connected Cars—Geschäftsmodelle*, S. 2–15. Retrieved May 11, 2021, from, https://www.bvdw.org/fileadmin/bvdw/upload/publikationen/digitale_transformation/Diskussionspapier_Connected_Cars_Geschaeftsmodelle.pdf

Capgemini. (2016). *Studie IT-trends 2016. Digitalisierung ohne Innovation?* S. 3–46. Retrieved April 4, 2021, from, https://www.capgemini.com/de-de/wp-content/uploads/sites/5/2016/02/it-trends-studie-2016.pdf

Chen, Y., & Li, L. (2014). *Advances in intelligent vehicles.* Elsevier Academic Press.

Deloitte. (2016). *Autonomes Fahren in Deutschland—wie Kunden überzeugt werden*, S. 1–19. Retrieved February 25, 2021, from, https://www2.deloitte.com/de/de/pages/consumer-industrial-products/articles/autonomes-fahren-in-deutschland.html

Deloitte II (2017): The race to autonomous driving. Winning American consumers' trust. In: Deloitte Review (20), S. 1–22. Online verfügbar unter https://www2.deloitte.com/content/dam/insights/us/articles/3565_Race-to-autonomous-driving/DR20_The%20race%20to%20autonomous%20driving_reprint.pdf. Accessed: 06.05.2021

Deloitte III. (2017). *What's ahead for fully autonomous driving—Consumer opinions on advanced vehicle technology. Deloitte global automotive consumer study*, S. 2–15. Retrieved May 6, 2021, from, https://www2.deloitte.com/content/

dam/Deloitte/us/Documents/manufacturing/us-manufacturing-whats-ahead-for-fully-autonomous-driving.pdf

Diehlmann, J., & Häcker, J. Dr. (2013). *Automotive management. Navigating the next decade* (2. Aufl.). Oldenbourg Verlag.

Diez, W. (2018). *Wohin steuert die deutsche Automobilindustrie?* (2. Aufl.). Walter de Gruyter GmbH.

EY II. (2016). *How much human do we need in a car? The evolution of artificial intelligence and the acceptance of autonomous vehicles*, S. 1–8. Retrieved April 18, 2021, from, https://www.ey.com/Publication/vwLUAssets/ey-autonomous-vehicles.pdf/%24FILE/ey-autonomous-vehicles-tl-report.pdf

Gerdes, C., & Maurer, M. (2015). *Autonomous driving. Technical, legal and social aspects*. Springer Verlag GmbH.

Goodyear, London School of Economics II. (2015). Millennial's views on the future of mobility in Europe. In *Think Good Mobility*, S. 4–24. Retrieved March 9, 2021, from, https://drive.google.com/file/d/0B1HvJzTnvhLfc0dOYWJtTnBfUTA/view

Gutzmer, A. (2018). *Marken in der Smart City. Wie die Cyber-Urbanisierung das Marketing verändert*. Springer Fachmedien GmbH.

Helmond, M., & Terry, B. (2016). *Lieferantenmanagement 2030. Wertschöpfung und Sicherung der Wettbewerbsfähigkeit in digitalen und globalen Märkten*. Springer Fachmedien.

Herrmann, A., Brenner, W., & Stadler, R. (2018). *Autonomous driving. How will the driverless revolution change the world*. Emerald Publishing Ltd.

Horváth & Partners, Fraunhofer. (2016). *The value of time. Nutzerbezogene Service-Potentiale durch autonomes Fahren*, S. 1–38. Retrieved May 6, 2021, from, https://www.horvath-partners.com/fileadmin/horvath-partners.com/assets/05_Media_Center/PDFs/deutsch/Studie_Value_of_Time_2016-04-22_FINAL.pdf

Hülsmann, M., & Fornahl, D. (2014). *Evolutionary paths towards the mobility patterns of the future*. Springer Verlag.

Humboldt-Universität zu Berlin, Geographisches Institut. (2016). User perspectives on autonomous driving. A use-case-driven study in Germany. In *Arbeitsberichte* (187), S. 3–96. Retrieved June 2, 2021, from, https://www.geographie.hu-berlin.de/de/institut/publikationsreihen/arbeitsberichte/download/Arbeitsberichte_Heft_187.pdf

IBM Institute for Business Value. (2016). *A new relationship—People and cars. How consumers around the world want cars to fit their lives*, S. 1–21. Retrieved April 18, 2021, from, https://www.ibm.com/services/multimedia/A-new-relationship-Exec-Report-v3.pdf

Institute for Mobility Research. (2018). *Autonomous driving—The impact of vehicle automation on mobility behaviour*, S. 1–95. Retrieved June 26, 2021, from, https://www.researchgate.net/profile/Stefan_Trommer/publication/312374304_Autonomous_Driving_-_The_Impact_of_Vehicle_Automation_on_Mobility_Behaviour/links/5881fe354585150dde4014a0/Autonomous-Driving-The-Impact-of-Vehicle-Automation-on-Mobility-Behaviour.pdf?origin=publication_detail

Karls, I., & Mueck, M. (2018). *Networking vehicles to everything. Evolving automotive solutions*. Walter de Gruyter GmbH.

Kearney, A. T. (2016). *How automakers can survive the self-driving era. A.T. Kearney study reveals new insights on who will take the pole position in the $560 billion autonomous driving race*, S. 1–36. Retrieved March 16, 2021, from, https://www.kearney.com/automotive/article?/a/how-automakers-can-survive-the-self-driving-era

Kosch, T., Schroth, C., Strassberger, M., & Bechler, M. (N.A.). *Automotive internetworking*. John Wiley & Sons.

KPMG II. (2017). *Global automotive executive survey 2017*, S. 1–56. Retrieved May 6, 2021, from, https://assets.kpmg.com/content/dam/kpmg/es/pdf/2017/global-automotive-executive-survey-2017.pdf

KPMG I. (2018). *Autonomous vehicle readiness index. assessing countries' openness and preparedness for autonomous vehicles*, S. 1–60. Retrieved July 22, 2021, from, https://assets.kpmg.com/content/dam/kpmg/nl/pdf/2018/sector/automotive/autonomous-vehicles-readiness-index.pdf

Kreutzer, R., Neugebauer, T., & Pattloch, A. (2017). *Digital business leadership. Digital transformation, business model innovation, agile organization, change management*. Springer Fachmedien GmbH.

L.E.K. (2017). *Mapping the road to autonomous vehicles. Executive insights*, S. 1–6. Retrieved April 2, 2021, from, https://www.lek.com/sites/default/files/insights/pdf-attachments/1958_Future_of_Autonomous_Cars_LEK_Executive_Inisghts.pdf

McKinsey & Company. (2014). Connected car, automotive value chain unbound. In *Advanced Industries*, S. 7–50. Retrieved June 2, 2021, from, https://www.mckinsey.de/files/mck_connected_car_report.pdf

McKinsey & Company I. (2015). *Ten ways autonomous driving could redefine the automotive world*, S. 1–4. Retrieved March 2, 2021, from, https://www.mckinsey.com/industries/automotive-and-assembly/our-insights/ten-ways-autonomous-driving-could-redefine-the-automotive-world

McKinsey & Company II. (2015). *Urban mobility at a tipping point*, S. 3–22. Retrieved May 6, 2021, from, https://www.mckinsey.com/business-functions/

sustainability-and-resource-productivity/our-insights/urban-mobility-at-a-tipping-point

McKinsey & Company III. (2015). Wettlauf um den vernetzten Kunden. Überblick zu den Chancen aus Fahrzeugvernetzung und Automatisierung. In *Advanced industries*, S. 6–44. Retrieved April 6, 2021, from, https://www.mckinsey.de/files/mckinsey-connected-customer_deutsch.pdf

McKinsey & Company II. (2016). *Car data: paving the way to value-creating mobility. Perspectives on a new automotive business model*, S. 5–21. Retrieved June 2, 2021, from, https://www.mckinsey.de/files/mckinsey_car_data_march_2016.pdf

McKinsey & Company III. (2016). Monetizing car data. New service business opportunities to create new customer benefits. In *Advanced industries*, S. 3–57. Retrieved February 22, 2021, from, https://www.mckinsey.com/-/media/McKinsey/Industries/Automotive%20and%20Assembly/Our%20Insights/Monetizing%20car%20data/Monetizing-car-data.ashx

McKinsey & Company, Bloomberg. (2016). *An integrated perspective on the future of mobility*, S. 5–62. Retrieved April 27, 2021, from, https://www.bbhub.io/bnef/sites/4/2016/10/BNEF_McKinsey_The-Future-of-Mobility_11-10-16.pdf

Meyer, G., & Shaheen, S. (2017). *Disrupting mobility. Impacts of sharing economy and innovative transportation on cities*. Springer International Publishing AG.

National Highway Traffic Safety Administration. (N.A.). *Automated vehicles for safety*. Retrieved July 15, 2021, from, https://www.nhtsa.gov/technology-innovation/automated-vehicles-safety

OECD/International Transport Forum. (2015). *Automated and autonomous driving. Regulation under uncertainty*, S. 1–32. Retrieved April 27, 2021, from, https://www.itf-oecd.org/sites/default/files/docs/15cpb_autonomous-driving.pdf

Qusay, H., Khan, A., & Madani, S. (2018). *Internet of things. Challenges, advances and applications*. Taylor & Francis Group, LLC.

Roland Berger. (2011). *Automotive landscape 2025: Opportunities and challenges ahead*, S. 1–47. Retrieved July 12, 2021, from, http://www.forum-elektromobilitaet.ch/fileadmin/DATA_Forum/Publikationen/Roland_Berger_2011_Automotive_Landscape_2025_E_20110228.pdf

Roland Berger, fka Forschungsgesellschaft. (2016). *Index "Automatisierte Fahrzeuge,"* S. 2–18. Retrieved May 11, 2021, from, https://www.rolandberger.com/publications/publication_pdf/roland_berger_index_autonomous_driving_q3_2016_final_d.pdf

PA Consulting. (2017). *Autonomous vehicles—what are the roadblocks*, S.1–16. Retrieved June 6, 2021, from, https://www.paconsulting.com/insights/2017/autonomous-vehicles/

PWC. (2018). *Five trends transforming the automotive industry*, S. 1–48. Retrieved April 9, 2021, from, https://www.pwc.at/de/publikationen/branchen-und-wirtschaftsstudien/eascy-five-trends-transforming-the-automotive-industry_2018.pdf

PWC, NTT Data Deutschland. (2013). *Automotive Retail—Die Zukunft beginnt jetzt*, S. 12–56. Retrieved June 6, 2021, from, https://www.pwc-wissen.de/pwc/de/shop/publikationen/Automotive+Retail+-+Die+Zukunft+beginnt+jetzt!/?card=13001

PWC - Strategy&. (2016). *Connected car report 2016. Opportunities, risk, and turmoil on the road to autonomous vehicles*, S. 5–63. Retrieved April 27, 2021, from, https://www.strategyand.pwc.com/reports/connected-car-2016-study

PWC - Strategy&. (2017). *The 2017 strategy& digital auto report. Fast and furious: Why making money in the "roboconomy" is getting harder*, S. 1–41. Retrieved April 27, 2021, from, https://www.strategyand.pwc.com/reports/fast-and-furious

Seipp, V., Michel, A., & Siegfried, P. (2020). Review of international supply chain risk within banking regulations in Asia, US and EU including proposals to improve cost efficiency by meeting regulatory compliance. *Journal Financial Risk Management (JFRM)*. https://doi.org/10.4236/jfrm.2020.93013

Siegfried, P. (2020). *Marketing- and Vertriebskonzepte von erfolgreichen Unternehmen—Fallstudien*. Akademische Verlagsgemeinschaft.

Skilton, M. (2016). *Building digital ecosystem architectures. A guide to enterprise architecting digital technologies in the digital enterprise* (1. Aufl.) Palgrave Macmillan.

The Boston Consulting Group, World Economic Forum. (2016). *Self-driving vehicles, robot-taxis, and the urban mobility revolution*, S. 3–26. Retrieved March 29, 2021, from, https://www.bcg.com/…/automotive-public-sector-self-driving-vehicles-robo-taxis-urban-mobility-revolution.aspx

Winkelhake, U. (2017). *Die digitale Transformation der Automobilindustrie. Treiber—Roadmap -Praxis*. Springer Verlag GmbH.

Wyman, O. (2017). *The Oliver Wyman automotive manager*, S. 1–61. Retrieved April 23, 2021, from, www.oliverwyman.de/content/dam/oliver-wyman/v2/publications/2017/jun/OliverWyman_AutomotiveManager2017_web.pdf

5

Business Model 2030: A Metamorphosis of the Automotive Landscape

5.1 Development Scenarios

5.1.1 Creation of the Smart City

Since the invention of engine-driven locomotive vehicles in the late nineteenth century, mobility has become an integral part of society. Privately owned cars, especially in Germany where the automotive manufacturing industry is enrooted historically and significantly contributes to the overall GDP, have been considered a prestige object, expressing the individual financial status and affiliation to a certain social class. But globally observable trends have triggered a rethinking of the way, personal mobility and privately owned cars are perceived. By 2030, roughly 60% of the world's population will be situated in cities, thus changing the population slope between urban and rural regions.[1] This phenomenon is enforced by an emerging middle class in threshold and economically expanding countries such as the BRIC states.[2] The existing infrastructure, however, is not

---

[1] cf. Diehlmann, Jens; Häcker, Dr. Joachim (2013) Automotive Management S.44ff.
[2] cf. Roland Berger (2011) Automotive landscape 2025 S.11ff.

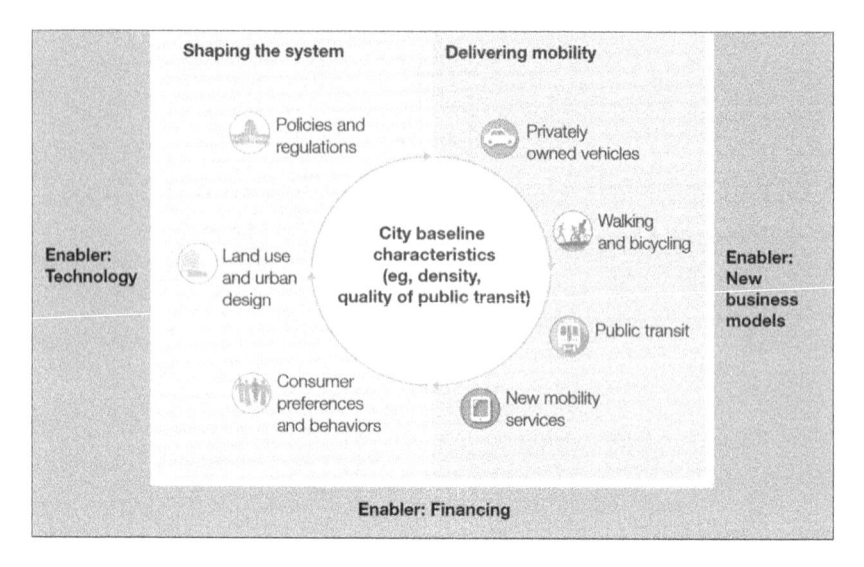

Fig. 5.1 A framework for understanding urban mobility. McKinsey & Company II (2015) S.4

construed to support such an increase in vehicles that participate in urban traffic. Already today, the congestion and its accompanying effects, e.g., lost time and wasted fuel can diminish the national GDP by up to 4%, since inefficiencies in transport systems are increasing the cost of doing business.[3]

The conversion of the city into a multi-modal smart city is dependent on a collaboration between governmental as well as municipal policy-makers that are shaping the blueprint for the urban infrastructure and companies providing new business models for delivering advanced mobility forms (Fig. 5.1).[4]

Regulators and governmental institutions are setting the framework in which municipal officials and, in a general view, mobility providers are operating to create the most valuable mobility ecosystem for the citizens and other users.[5] Defining suited traffic regulations laws and standards

---

[3] cf. Nieuwenhuis, Paul; Wells, Peter (2015) The Global Automotive Industry S.153–156.
[4] cf. Meyer, Gereon; Shaheen, Susan (2017) Disrupting Mobility S.149.
[5] cf. Roland Berger II (2016) A CEO agenda for the (r)evolution of the automotive ecosystem.

regarding the collection, assessment and distribution of car and personal data as well as regulating controversial topics such as technical certification of the connected vehicles, data ownership rights, and intellectual property rights are a premise to develop an adequate mobility environment.[6] Municipals can, in reconciliation with OEMs and mobility service providers, design the urban outlay regarding land use for public transport systems, vehicle-based traffic, and complementary locomotive means. The challenge for OEMs will be to capture a profitable role in the mobility ecosystem[7] by either providing the vehicle fleets necessary for various mobility solutions such as autonomous taxis or car sharing companies or by engaging and investing in mobility platforms that combine intermodal transport systems.[8] The triumvirate of entities developing advanced technologies that enable V2X communication,[9] providing the financial budget to implement vehicle and infrastructure-related adjustments as well as commercial players seeking to create new business models supporting the re-design of urban landscape, is essential to foster the smart city.

In a long-term vision, municipal and governmental officials are expecting an improvement of the external effects that individual automotive mobility has on today's society and environment.[10] One of the major drawbacks of the global penetration of privately-owned vehicles is the increasing pollution induced by combustion engines and the resulting emission of carbon and nitrogen dioxide. The exhaust scandal that was revealed in 2015, raised further awareness about the already controversially discussed use of otto-cycle and diesel engines and, thus, led to governmental regulations, discouraging the purchase of unsustainable vehicles. Automotive producers are challenged to adhere to stricter emission threshold values to avoid higher taxes on new models and derivatives

---

[6] cf. Diehlmann, Jens; Häcker, Dr. Joachim (2013) Automotive Management S.218.
[7] cf. BearingPoint Institute II (2017) Re-thinking the European Business Model Portfolio for the Digital Age S.7.
[8] cf. McKinsey & Company; Bloomberg (2016): An integrated perspective on the future of mobility S.17.
[9] cf. Meyer, Gereon; Beiker, Sven (2018) Road Vehicle Automation 5 S.91.
[10] cf. Deloitte University Press (2015) Patterns of Disruption S.14–16.

and are forced to publish fuel economy data according to the newly introduced Worldwide harmonized Light Vehicle Test Procedure (WLTP). This testing procedure leads to a more realistic indication of fuel consumption data and new taxation models for inefficient combustion engines.[11] The governmental target to increase the other transport modes in terms of convenience is simultaneously fostered by dedicating more space to walking, cycling, and public transport systems. New mobility models such as car sharing, e-mobility, and bike sharing are supported with subsidies and dedicated parking spaces to increase their attractiveness in congested urban areas.[12]

A major factor that needs to be resolved in order to comprehensively cover all city areas and grant all inhabitants access to the smart infrastructure is the so-called last-mile problem.[13] In the graphic below, the annual costs for mobility in San Francisco are depicted. Although this comparison of total costs of ownership for various mobility modes is limited to one city, it gives insight into a problem that is transferable to all large cities worldwide. For commuters living in suburbs or rural areas, financing a new or used car is favorable if the yearly number of miles driven exceeds—in this example—7000 or respectively 10,000 (Fig. 5.2).

Even assuming that new mobility services such as Electronic Hailing (e-hailing) or car sharing were accessible outside city centers, a majority of commuters would still benefit financially from owning a private car. Public transport, as a cost-efficient alternative, is challenged with overcoming the last-mile problem: The connection from people's home to the public transport system and analog.[14] Expanding public transport networks by adding lines, increasing capacities and frequencies could reinforce its attractiveness for commuters in terms of time efficiency and flexibility. Due to low utilization rates, the particular challenge for municipal planners is the financial support of an expansion of the public transport systems.

---

[11] cf. Waschl, Harald; Kolmanovsky, Ilya (2018) Control Strategies for Advanced Driver Assistance Systems and Autonomous Driving Functions S.41.
[12] cf. Gerdes, Christian; Maurer, Markus (2015) Autonomous Driving S.387.
[13] cf. Cordon, Carlos et al. (2016) Strategy is Digital S.85.
[14] cf. Vogel, Hans-Josef; Weißer, Karlheinz; Hartmann, Wolf D. (2018) Smart City: Digitalisierung in Stadt und Land S.34–35.

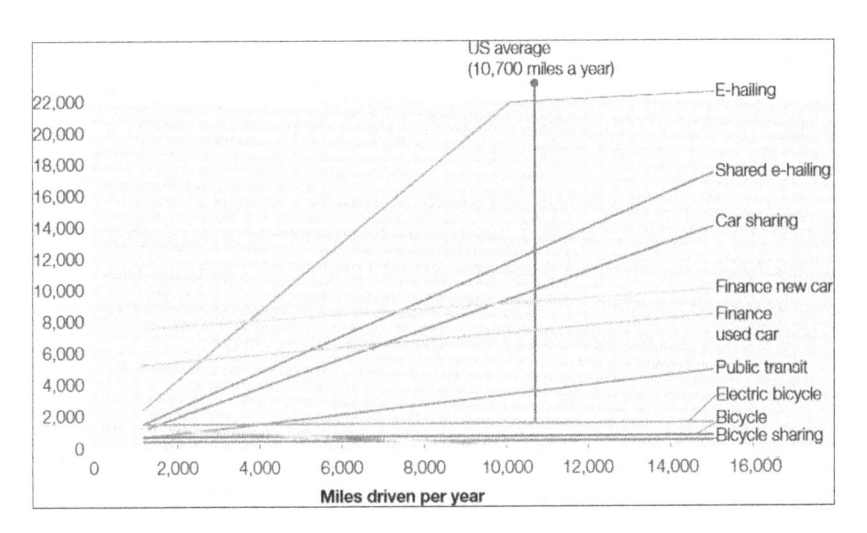

5 Business Model 2030: A Metamorphosis of the Automotive...

Fig. 5.2 Annual cost of mobility in San Francisco Bay Area (in US$). McKinsey & Company II (2015) S.9

The introduction of self-driving vehicles, in a long-term vision, leads to many societal benefits such as equitable access to low-cost mobility, decrease in pollution and traffic-related accidents, or improved customer service.[15] Municipal planners and governmental representatives are, however, confronted with a conflict of objectives, a trade-off between risks and rewards of a reconfiguration of the infrastructure towards a smart city. Given that robo-taxis or affordable privately owned autonomous vehicles will represent a convenient alternative to public transport networks, the costs for maintaining and reconstructing municipal transport systems could exceed the respective earnings.[16]

Furthermore, a wide-spread access to SDVs would allow citizens to move to suburban areas since the cost for commuting could still be favorable in comparison to higher rents in the city-center which would undermine current urban design plans. Typically, cities generate revenues from parking tickets, fuel taxes, or further vehicle-related fees and charges. Autonomous vehicles could cut these revenue streams unless other

---

[15] cf. Anderson, James et al. (2016) Autonomous Vehicle Technology S.17–22.
[16] cf. Jackel, Michael (2015) Smart City wird Realität S.238–241.

alternatives are introduced to support the city treasury such as taxation of road usage or the implementation of charging stations for electrified vehicles.[17] In general, all these considerations are underpinned by the same question, namely how to sustainably fund the virtual and physical infrastructure that is needed to create an ecosystem[18] that fosters social benefits and is supported by a new public–private funding scheme. The subsequent sections will present various smart city mobility models and key trends that affect the way the infrastructure has to be configured in order to maximize the economic potential that these mobility concepts promise.

V2X communication represents the ability of the car to communicate with its exterior by being integrated into the Internet of Things and, thus, creates the premise for the development and establishment of new mobility models that are dependent on a boundless integration.[19] Customers and all mobility stakeholders need to have access to usage data either via cloud services or provided directly from OEMs that publish on-board stored information on data platforms (Fig. 5.3).

The synchronization of accumulated and real-time data is essential for the efficient configuration of car sharing or e-hailing solutions and depicts an indispensable prerequisite for the introduction of self-driving vehicles. Humans are increasingly replaced as the sole source of intelligence in operating the vehicle. The shifting interface between the user and his car towards a computer-based interaction requires the installation of radar and interface capabilities[20] provided by satellite and drone service providers as well as computing centers that, based on big data analytics, support the continuous improvement of machine learning and autonomous driving capacities. Every vehicle will be equipped with one central data system with interfaces for OEMs, car sensors, other traffic participants, the driver, local hubs, satellites, and mobile networks. Advanced sensors that receive and transmit precise information as well as transmission

---

[17] cf. Neckermann, Lukas (2015) The Mobility Revolution S.50–53.
[18] cf. OECD/International Transport Forum (2015) Automated and Autonomous Driving S.18.
[19] cf. Hassan, Qusay (2018) Internet of Things - A to Z S.10ff.
[20] cf. Herrmann, Andreas; Brenner, Walter; Stadler, Rupert (2018) Autonomous Driving S.62–66.

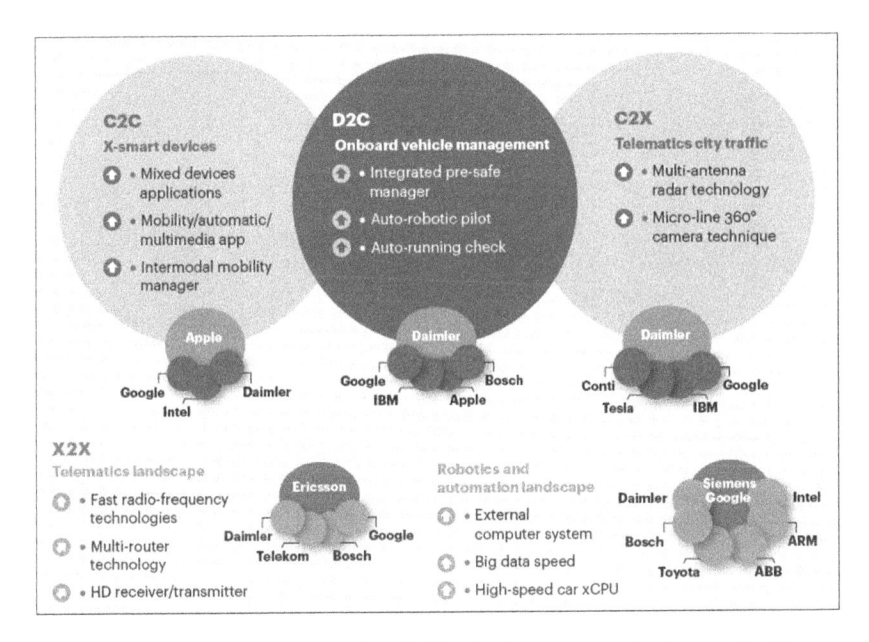

5 Business Model 2030: A Metamorphosis of the Automotive…

Fig. 5.3 Strategic alliances in the field of V2X communication. A.T. Kearney (2016) S.22

technologies will have to bear situations in which the car encounters objects such as other cars, bicyclists, pedestrians, and man-made barriers.[21]

Since autonomous vehicles are subject to stricter safety regulations and testing procedures, the information that sensors must collect and process have to cover its entire surroundings to be able to calculate any possible traffic and driving situation. For an efficient construction of the smart city, the comprehensive usage of C2X communication represents a complex hurdle. An intermodal mobility infrastructure will be based on sustainable measures and dashboards that allow for a real-time tracing of traffic management issues or the recalibration of municipal mobility services (Fig. 5.4).[22]

The control of multiple mobility providers and city operations requires a constant exchange of information flows between all stakeholders that

---

[21] cf. Karls, Ingolf; Mueck, Markus (2018) Networking Vehicles to Everything.
[22] cf. Linnhoff-Popien, Claudia; Schneider, Ralf (2018) Digital Marketplace unleashed S.407ff.

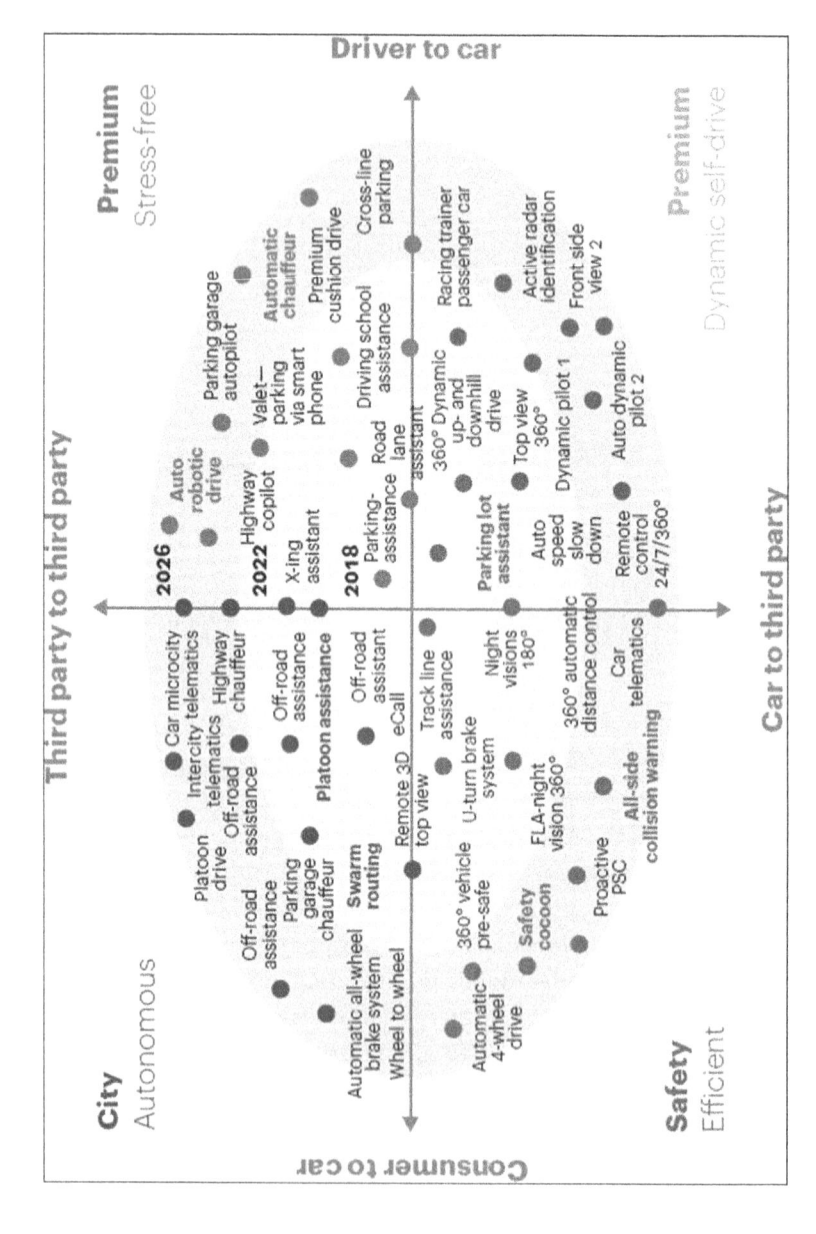

Fig. 5.4 Connectivity features will evolve along four dimensions. A.T. Kearney (2016) S.31

are involved in the infrastructure and mobility services. Advanced monitoring systems for accurate reporting on road capacity utilization, traffic flows, and maintenance needs of public infrastructure are essential to optimize the complex transportation and logistics streams.[23] Since autonomous driving between different telematics and online systems adds a thick layer of complexity, a new level of collaborative effort among governmental entities and industry players is essential to create an infrastructure that enables a 360° steering of internal and external intelligence systems.

Centralized and decentralized hotspots with local IT applications and mobile broadcasting systems are required to interconnect with stationary networks or in-car hotspots.[24] The automotive as well as the telecommunication industry need to collaboratively create the technological requirements that enable the creation of the smart city and the integration of new mobility business models into the new infrastructure. To encounter this challenge, automotive companies are building strategic alliances and investing in novel mobility services to gain a control point in the future traffic landscape exploiting emerging revenue pools and streams.[25]

New and improved mobility services, as already depicted in Chap. 3, are continuously increasing consumer choice options and mobility convenience in terms of availability, transportation speed, and flexibility. Furthermore, mobility is becoming multimodal, on-demand, and shared with the consequence that the overall transportation system including fleet management and traffic control centers are getting digitalized in order to efficiently match demand and supply in the short and mid-term.[26] Already today, several cities such as Barcelona, Boston, and Rio de Janeiro have partnered with Waze, a USA-based start-up that developed a crowd-sourcing mapping application, to integrate the data into the city's intelligent transportation system traffic control center.[27] This enables drivers to receive detailed, user-generated real-time data to avoid bottle-

---

[23] cf. Meier, Andreas; Portmann, Edy (2016) Smart City S.173.
[24] cf. Kreutzer, Ralf; Neugebauer, Tim; Pattloch, Annette (2017) Digital Business Leadership S.8ff.
[25] cf. Siegfried, Patrick; Zhang, John Jiyuan (2021) Developing a sustainable concept fort he urban lastmile deivery S.5.
[26] cf. KPMG III (2017) Islands of AutonomyS.18–20.
[27] cf. McKinsey & Company V (2016) Urban mobility 2030 S.12–16.

necks while the respective municipal traffic authorities can use the information on traffic conditions to respond to emerging situations. In general, the success of novel mobility services is highly dependent on two critical factors: costs and convenience. Automotive manufacturers as well as market newcomers from the telecommunication or software industry have to configurate their mobility services to best match those key criteria. The subsequent section depicts various mobility concepts and gives examples of different services, already implemented in cities all over the world.

In the past several years, e-hailing services, especially the USA-based service provider Uber[28] have triggered a controversial discussion in Germany about possible impacts on the local taxi industry and the associated loss of workplaces. While Uber has been widely restricted in Germany, the former start-up managed to introduce its services in roughly 300 cities in 58 countries with a strong tendency to rise. The Chinese market, in particular, has shown exponential growth rates leading to around 170 million registered users of e-hailing services.

Uber has confirmed to enter a strategic partnership with the Carnegie Mellon University Robotics Institute to co-develop self-driving cars which could even lead to a more personalized mobility experience. The emergence of car sharing options is considered a backlash on the changing attitude of mostly metropolitan inhabitants, favoring flexible and shared mobility services over privately owned vehicles that are associated with a high total cost of ownership and low utilization rates, especially in urban areas.[29] The industry is observing double-digit growth rates, exceeding the one million user threshold in Germany already in 2014. Car manufacturers have geared up for this revolution by launching their own sharing services across German metropolitan areas. Daimler's car2go and BMW's DriveNow are even in the process of merging to further expand their market penetration and increase the accessibility for its users. Since shared vehicles have less idle periods and get used more intensely, the increase in annual mileage and capacity utilization contributes to ease the congestion problems and optimize the traffic flow.[30]

---

[28] cf. KPMG I (2016) I see. I think. I drive. (I learn)
[29] cf. Deutsche Bank AG (2017) The digital car S.25–28.
[30] cf. BearingPoint; IIHD (2017) Ecosysteme & Plattformen verändern die Handelslandschaft.S.10ff.

Estimates assume that the comprehensive use of such services could lower the cost of personal mobility by 30–60% relative to private car ownership. Company-driven, on-demand shuttles and private busses have already been implemented by private employers such as Google, Apple, or Genentech.[31] These new crops of connected, on-demand transportation networks are installed to relieve the employees from rush-hour traffic and congested roads and eventually contribute to their working morale and performance. For most city dwellers and commuters, public transport still represents the most used and economic mode of locomotion. Municipal institutions worldwide have recognized the pressure that the increasing traffic density is putting on the urban public infrastructure. Bogota with its TransMilenio Bus rapid-system (BRT) has tripled their capacity to 241 miles with dedicated bus lanes, elevated bus stations, and a smart-card payment system that eases the use and charging of services. Dubai's roads and transport authority has introduced an urban master plan that includes 262 miles of newly installed metro lines by 2030 and San Francisco endeavors to extend its Caltrain to reach the city center,[32] upgrading the commuter rail system and providing bus rapid transport along select corridors.

The most promoted feature of smart cities is the combination and linkage of all existing mobility modes in an intermodal approach to creating a seamless urban connection. Prerequisites of consolidating all modes of transport and integrating new mobility on-demand services are the digitization of the public transport system and the creation of a mobility ecosystem that is accessible via a comprehensive platform model. Helsinki is currently developing an on-demand mobility program whose target is to incrementally decrease personal car usage by 2025. This mobility as-a-service concept is based on a mobile app that enables the user to book and pay in one click for any trip by train, taxi, bus, bike, or shared car. The installed platform follows a single-ticket principle, a so-called mobility ticket,[33] that embraces all booked modes of transport. The Californian

---

[31] cf. EY (2014) Deploying autonomous vehicles S.4–7.
[32] cf. The Boston Consulting Group; World Economic Forum (2016) Self-driving vehicles, robot-taxis, and the urban mobility revolution S.3.
[33] cf. Roland Berger (2011) Automotive landscape 2025 S.17–19.

Valley Transportation Authority (VTA) has installed innovation hubs where private companies, start-ups, and students are encouraged to co-develop new transportation-related technology that is laying the groundwork for intermodal transportation systems in which SDVs and robo-taxis are an integral element, complementing the publicly offered transport modes.[34] Moovel, a subsidiary of Volkswagen developed a planning application that integrates several user options including public and private mobility services are already available in five regions in Germany. The configuration of a smart city thus requires the creation of an own "micro-mobility center"[35] that demands a high degree of compatibility and integration of traffic information, including personal and financial data of its users, not only within the own city center but also across different connecting smart cities. Aside from automotive manufacturers, governments are also confronted with identifying new revenue pools and accessing new revenue streams to support the installation and future enhancements of the new infrastructure.

Municipal authorities need to collect taxes or tolls while service providers from all industries ranging from telecommunication, insurance, entertainment, or the software sector will want to track usage and charge for service offerings,[36] all in real-time via flat rates, subscriptions, or a payment bonus accrued through usage and other loyalty programs (Fig. 5.5).

The graphic above illustrates the target picture of a municipal traffic management system with all relevant features and IT-based information and communication networks. The core areas in which OEMs need to intervene in order to safeguard the future business volume and establish new business models that can be roughly grouped into three clusters. Beyond the traditional sale of vehicles with different powertrains to private persons, car and bus fleets are needed to support the concept of new mobility service providers such as Uber, Lyft, or company-owned

---

[34] cf. The Boston Consulting Group; World Economic Forum (2016) Self-driving vehicles, robot-taxis, and the urban mobility revolution S.3–5.
[35] cf. Deloitte I (2017) The Future of the Automotive Value Chain S.35–39.
[36] cf. Vogel, Hans-Josef; Weißer, Karlheinz; Hartmann, Wolf D. (2018) Smart City: Digitalisierung in Stadt und Land S.63.

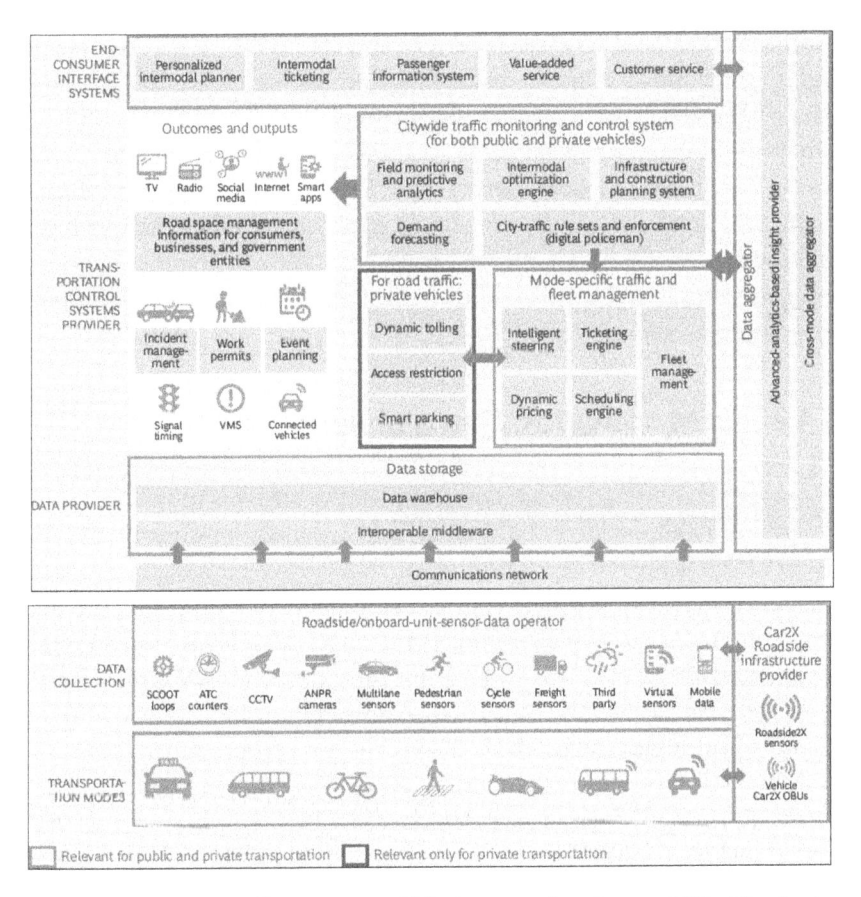

5 Business Model 2030: A Metamorphosis of the Automotive... 107

Fig. 5.5 Municipal Traffic Systems will help control and steer SDVs. The Boston Consulting Group/World Economic Forum (2016) S.16

transport modes.[37] As depicted in the upcoming sections about different development scenarios, most countries will still highly depend on individual car utilization since only a minor part of wealthy, industrialized nations with a high-income population will already have created an environment in which the sale of autonomous vehicles can be fostered profitably.[38] But OEMs have realized the importance to engage in two further

---

[37] cf. McKinsey & Company II (2015) Urban mobility at a tipping point S.15–17.
[38] cf. Morgan Stanley (2012) Global Auto Scenarios 2022.

promising business activities. The car and its hardware will represent a significant control point in collecting and aggregating onboard sensor data. Advanced analytics-based insights into the car utilization and customer profiles are highly valuable to access and safeguard future revenue pools. Furthermore, most German OEMs are already competing with high-tech companies and start-ups to establish mobility platforms and end-consumer interface systems[39] by promoting car sharing services or integrated mobility apps and, thus, installing self-controlled mobility ecosystems as referred to in Chap. 3.

5.1.2 OEM Scenarios: Forms of Repositioning

There are three scenarios that describe how the business model, corporate design, and value-added chain of OEMs are likely to alter in the next decade. The degree of repositioning of automotive manufacturers in the market is thereby dependent on their willingness to change and the business potential that emanates from the change. Transformation strategies for mass market producers will significantly deviate from those of niche producers such as Ferrari, Bugatti, or Lamborghini since sportscar manufacturers are highly focused on delivering a technical superior product and reach far less market penetration.[40]

The classic OEM, as already depicted in Chap. 2, is oriented to traditional business models and value-added chains with a clear focus on product and technology-related core competencies. Vertical supply chains with different tier-n suppliers as well as in-house development of technical vehicle components and features coin the primary business activities.[41] Since passenger vehicles are a complex industrial product, the core competencies in the technical and engineering function represent the major competitive advantage.

Corporate reconfigurations are narrowed to the optimization of established processes and structures without essential innovations in the

---

[39] cf. Simpson, Timothy; Siddique, Zahed; Jiao, Jianxin (2006) Product Platform and Product Family Design S.49–52.
[40] cf. Natalia, Nefedova (2013) Trends in Automotive Industry S.15–17.
[41] cf. Oswald, Gerhard; Krcmar, Helmut (2018) Digitale Transformation S.13ff.

market-ready vehicle models and derivatives. There is also a strict separation in the allocation of competencies between the OEM and co-operation partners from the information and communication technology industry. The complementary character of these contractual partners represents a win-win situation for both parties since OEMs benefit from the partner's digital expertise while ICT companies gain insight into the technical production procedure.

The so-called 180° OEM complements its traditional business model with ICT partnerships that contribute to the development of single digital services and, thus, have a selective impact on the digital value-added chain.[42] Automotive manufacturers in this category pursue new approaches that are not limited to classical vehicle features and characteristics but partially expand their product portfolio. ICT partnerships or the acquisition of start-ups in the ICT industry aid in the development, implementation, and commercialization of technological add-ons and digital services and, thus, increasing the in-house generated influence on the digital value chain.[43] This results in a strengthening of the customer relation and interaction and facilitates the companionship of the customers in most stages of their life-cycle. The synergy of decades-long engineering experience enriched by digital services ensues a solid competitive position.

With regard to 2030, most mass-market producers aspire to transform into a 360° OEM that successfully links the traditional and digital value-added chain and disposes of a comprehensive product and service portfolio of vehicle-dependent and independent offerings.[44] Technological and digital competence gained through strategic partnerships with or acquisition of ICT companies foster the development and establishment of ecosystems that incorporate the customers' requirements for networked services and automated functions.[45] The intelligent interconnection between the traditional and digital value chain increases the depth of added value and safeguards the provision of up-to-date, individual

---

[42] cf. Huber, Walter (2016) Industrie 4.0 in der Automobilproduktion S.118ff.
[43] cf. Diehlmann, Jens; Häcker, Dr. Joachim (2013) Automotive Management S.35.
[44] cf. Fisher, Tony (2009) The Data Asset S.61ff.
[45] cf. Roland Berger; Lazard (2016) Global Automotive Supplier Study S.25–28.

vehicle-dependent, and independent product features and services throughout the entire customer life cycle. OEMs are capable of retracing and analyzing the data track of their customers from the purchase decision up until their personal utilization habits. The implicit control of the single digital customer touchpoints thereby significantly improves the OEM's competitive advantage in enhancing product features and customer experience concepts.

5.1.3 Mobility Scenarios: Extension of the Mobility Landscape

For the subsequent evaluation of future market penetration scenarios of autonomous vehicles, the criteria of segmentation have to be reassessed. Traditional segmentation clusters, as addressed in Chap. 2, that are primarily based on customers' age, income, or residential area, are incrementally becoming obsolete in view of the altering value proposition and connected vehicles and emerging mobility concepts.[46] Today's business environment requires a more granular view of mobility markets and old segmentation criteria such as sales country or region are increasingly replaced by city characteristics and phenotypes. New models embrace population density, economic development, and prosperity as differentiation factors and forecast that consumer preferences, policy, and regulation as well as availability and pricing of new business models will significantly diverge across those segment clusters (Fig. 5.6).[47]

In metropolitan, high-income areas as London or Shanghai shared and on-demand mobility-as-a-service concepts present a competitive value proposition considering local congestion fees, lack of parking, or traffic jams. In contrast, in rural areas, where low density creates a barrier to scale, private car usage will most likely remain the preferred means of transport.[48] As illustrated in the graphic above, OEMs are confronted

---

[46] cf. Rosenzweig, Juan; Bartl, Michael (2015) The Making of Innovation S.4–8.
[47] cf. McKinsey & Company III (2017) The future(s) of mobility S.6–10.
[48] cf. KPMG I (2017) Reimagine places: Mobility as a Service S.17–19.

5 Business Model 2030: A Metamorphosis of the Automotive… 111

The effects of car sharing, urbanization, and macroeconomics on vehicle sales vary strongly by region and city type

HIGH-DISRUPTION SCENARIO

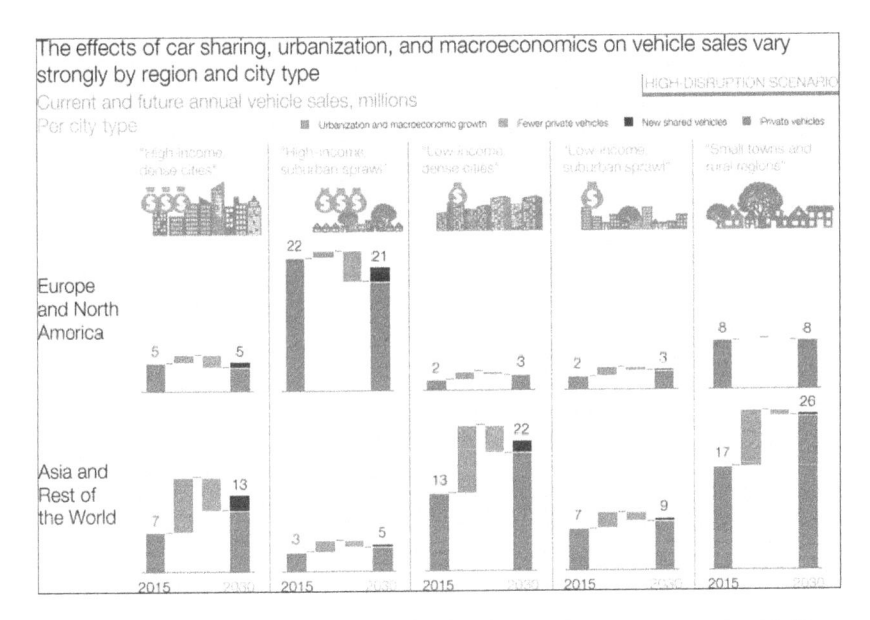

Current and future annual vehicle sales, millions
Per city type

Fig. 5.6 The impact of car sharing, urbanization, and macroeconomics. McKinsey & Company I (2016) S.10

with stagnating or even declining sales in the European and North American markets. The regressive private vehicle sales volume can be compensated for by focusing on providing car sharing fleets or other vehicles necessary for novel mobility services. In Asia, the BRIC states, and other economically emerging countries, traditional vehicle sales are predicted to rise significantly due to growing local middle classes and the increasing prosperity of social classes that have formerly been excluded from the private car market (Fig. 5.7).

Since fully autonomous vehicles are unlikely to be commercially available before 2020,[49] advanced driver assistance systems will take a bearing role in preparing regulation authorities, consumers, and eventually corporations in the medium-term until the autonomy disruption will

---

[49] cf. U.S. Department of Commerce, Economics and Statistics Administration (2017) The Employment Impact of Autonomous Vehicles S.4–5.

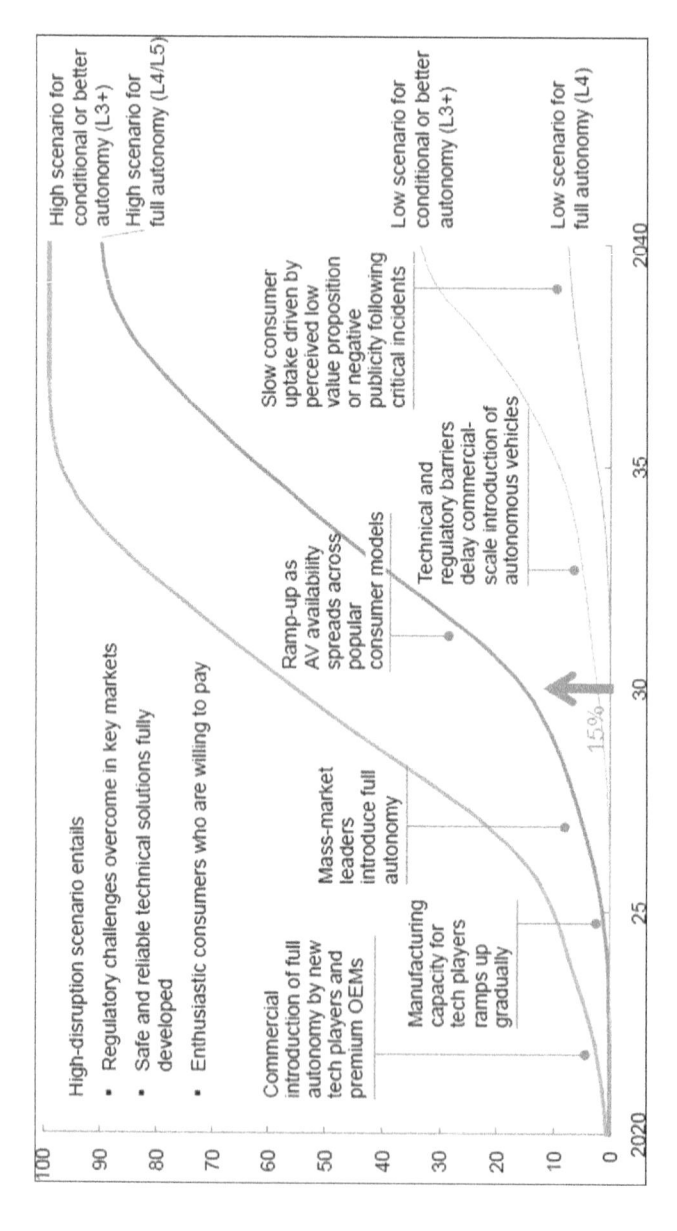

Fig. 5.7 Fully autonomous vehicle share of new vehicle market. McKinsey & Company I (2016) S.11

5 Business Model 2030: A Metamorphosis of the Automotive...

become reality. The illustration above shows potential low- to high-disruption scenarios in the time frame from 2020 to 2040. Relevant factors that impede a faster market penetration of ADAS features are pricing, consumer understanding, and safety issues, which directly affect the customers' willingness to pay and, thus, the sales volume. Furthermore, technological challenges will cause a delay between the introduction of conditionally autonomous vehicles that allow the driver to cede control in certain situations and fully autonomous cars requiring no driver intervention during the entire journey. Overcoming this technological hurdle between the level 3 and level 4 autonomy definition, determined by the National Highway Traffic Safety Administration, will be the crucial step in entirely establishing the automotive disruption.[50] In this scenario analysis, 50% of vehicles sold in 2030 could be highly autonomous while 15% could already be fully autonomous.

In the subsequent section, three models—clean and shared, private autonomy, and seamless mobility—will be introduced that show a diverging degree of autonomous vehicle penetration based on differing city type criteria. Delhi, Mexico City, or Mumbai can be adduced exemplarily when considering the first metropolitan phenotype. Densely populated metropolitan areas in developing countries that experience rapid urbanization and suffer from congestion and inferior air quality. Due to insufficient infrastructure, interferences from pedestrians, the variety of vehicles participating in traffic, and the lack of clear adherence to traffic regulations, widespread use of self-driving vehicles is not an option for the short- or medium-term. The most constructive approach to improve the status quo is an incentivized shift to cleaner transport modes such as electrified vehicles or plug-in-hybrids by limiting car ownership, optimizing shared mobility, and expanding public transit. The private autonomy concept refers to cities where municipal development and commuting patterns have increased urban sprawl significantly such as Los Angeles.[51] While there is no apparent necessity for private car ownership since public transit systems are installed sufficiently, privately used vehicles will still

---

[50] cf. OECD/International Transport Forum (2015) Automated and Autonomous Driving S.14.
[51] cf. The Boston Consulting Group; World Economic Forum (2016) Self-driving vehicles, robot-taxis, and the urban mobility revolution S.23–26.

represent the major share of transportation in the foreseeable future. But new mobility technologies such as SDVs and EVs will incrementally find their way into the cityscape. Advances in connectivity will ease the implementation of demand-driven congestion charges and foster an increase in road capacity utilization while limiting the need for new constructions. Car sharing and e-hailing solutions will emerge as a complementary option for city dwellers but will not replace private vehicle mobility on a large scale.[52] But the emergence of autonomous and electrified vehicles also has a drawback. Lower marginal costs to travel an extra mile in fuel- or energy-efficient cars paired with the convenience of autonomy could increase the demand for mobility. The trade-off, posed by this scenario, between using large-capacity buses and low-occupancy SDVs triggered by increased comfort and better end-to-end services could further add to the current congestion issue and increase the additional distance traveled by up to 50%.[53] Beyond that, autonomous mobility will address new customer segments such as children, elderly or disabled persons that aggravate the infrastructure capacity utilization.

The third and most radical concept compared to today's reality is seamless mobility which will primarily emerge in densely populated, high-income cities including Chicago, Hong Kong, London, or Singapore. The provision of mobility will be safeguarded by a combination of self-driving and shared vehicles with high-quality public transit networks as a backbone. Mobility services are predominantly door-to-door[54] and on-demand leading to blurred boundaries among private, shared, and public transport. Integrated platforms that manage multi-model traffic flows and deliver mobility-as-a-service will create a comprehensive mobility ecosystem accessible by all urban residents.

---

[52] cf. Ecola, Liisa et al. (2015) The Future of Mobility S.35.
[53] cf. Gerdes, Christian; Maurer, Markus (2015) Autonomous Driving S.180ff.
[54] cf. Vogel, Hans-Josef; Weißer, Karlheinz; Hartmann, Wolf D. (2018) Smart City: Digitalisierung in Stadt und Land S.34–36.

5.2 A New Appreciation of the Value Chain

5.2.1 Value Chain Integration: Traditional Versus Digital

The speed of innovations conquering the market has accelerated and competitive products and services are increasingly shaping more ambitious and challenging customer needs. In the past, the focal point of OEMs and suppliers has been the development and production of highly complex industrial products, driven by engineering and technical expertise as well as an established network of capital-intense processes fraught with risk.[55] Existing business models are incrementally influenced by digitalization and connectivity whereat sales volumes in classic vehicle segments is forecasted to decline or stagnate. However, the vehicle, as a source of data supply, facilitates the generation of unforeseen possibilities to accompany the customer during his entire life cycle with individual vehicle-dependent and -independent products and services.[56] A user-friendly configuration of central control elements and advanced tools of interaction with the customer contributes to maintaining control over data and consumers (Fig. 5.8).

A linkage of the traditional and digital value-added chains to expand the customer touchpoints is considered the key to success. Beyond increased user experience and insight, digital services have the potential to enlarge the revenue pool substantially since the average margins are higher.[57] Audi has even announced that its service portfolio will presumably account for 50% of its sales volume by 2020. To reach such ambitious goals, OEMs as well as suppliers will have to rebalance their corporate functions and organization forms.[58] The subsequent sections give insight into the requirements to maintain and introduce further business models along their new digital-oriented path.

---

[55] cf. The Boston Consulting Group (2004) Beyond Cost Reduction S.15–17.
[56] cf. Linnhoff-Popien, Claudia; Schneider, Ralf (2018) Digital Marketplace unleashed S.633ff.
[57] cf. Wedeniwski; Sebastian (2015) The Mobility Revolution in the Automotive Industry S.102–104.
[58] cf. Caudron, Jo; Peteghem, Dado (2018) Digital Transformation.

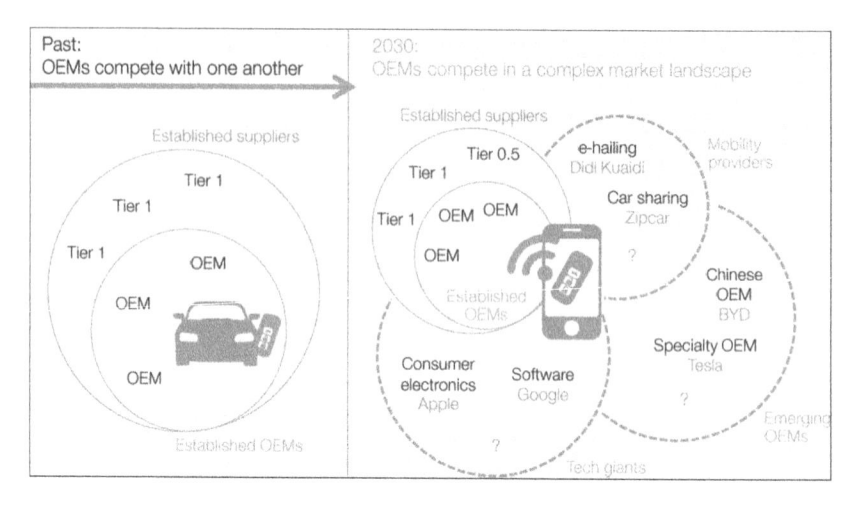

Fig. 5.8 Increasing complexity of the competitive landscape. McKinsey & Company I (2016) S.13

Shortening product life cycles and a dynamic, innovation-driven competitive environment are key challenges jeopardizing the way established corporations organize their functions.[59] Agile processes gain in importance since components with a high share of creativity such as software and parts with delimited interfaces and low safety relevance can be developed in a short-term time frame, within agile and fluid organization forms. The ability to accelerate the market introduction of features with short life cycles will be a critical factor in solving the so-called "clock-speed dilemma."[60] Safety-related components with high integration standards and regulatory specifications still require comprehensive testing and safeguarding measures within the development process. Therefore, classic components in the areas of powertrains, chassis frame, steering, passenger restraint, braking, and exhaust systems will still rely on the established, perennial product development process. However, a comparison of the development approaches between the automotive and IT industries shows that vehicles are, to a large extent, developed along a classic waterfall principle with structured release processes. The

---

[59] cf. Siegfried, Patrick (2015) Trendentwicklung und strategische Ausrichtung von KMUs S.87–95.
[60] cf. Oliver Wyman IV (2016) Sourcing in the automotive industry S.4–6.

increasing integration of software requires a recalibration of development paces to solve the clock-speed dilemma.[61] In the past, OEMs have neglected the asynchronism of development cycles between hardware and software by introducing new software features only in facelifted derivatives or new models. But dynamic product development with short-run feedback loops is essential to synchronize the cycle frequency with those of incumbent high-tech companies and start-ups since, otherwise, competitors will determine standards regarding software and user experience. OEMs need to expand their modular construction system to components that are relevant for digital business models. If these parts became exchangeable, automotive producers could provide advanced functionalities and access the second-hand car market, creating a new lever of after-market revenues.[62]

5.2.2 The Significance of Partnerships and Strategic Alliances

The setup of digital expertise and a skilled workforce that is capable of introducing and managing novel business models and digital services leads to a simple make-or-buy decision. OEMs have to determine which competencies are to be created in-house and which functions can be complemented by strategic partnerships with companies from the information and communication technology sector.[63] One of the primary challenges is the technological development and implementation of digital ecosystems regarding the infotainment system which represents the heart of the customer interface. In the past, OEMs heavily relied on distinct, in-house developed solutions and abstained from offering an open-source application programming interface. The end-to-end closeness of the system, in fact, deters competitors from interfering in the application business but it also keeps off innovative start-ups that could contribute to expanding the in-car service portfolio.[64] A promising solution for inte-

---

[61] cf. KPMG I (2018) Autonomous Vehicle Readiness Index 3.7ff.
[62] cf. Pollak, Dale (2017) Like I See It.
[63] cf. Diez, Willi (2018) Wohin steuert die deutsche Automobilindustrie? S.199ff.
[64] cf. Iskander Business Partner (2016) Digitalisierung in der Automobilindustrie S.14–17.

grating external innovation capacities into the process is the development of a central back-end platform for digital services accessible by multiple OEMs and external software developers. The open-source concept of such a system offers the chance to create a scalable ecosystem whose critical size and, thus, profitability is obtained quicker.

A collaboratively managed application store, similar to those of Apple or Samsung, facilitates the co-development of use cases and expanding the limited portfolio of applications. Incrementally increasing scale will, along the way, create network effects attracting external developers to participate in advancing the platform which, in turn, triggers new users.[65] This mutual incentive structure facilitates a quick scalability of the ecosystem. Furthermore, since resources needed to manage the platform are shared among the companies involved and boundaries between the industries are becoming blurred, platforms represent an asset-light business model, diversifying and reducing the individual risk regarding initial investments and running costs. But OEMs must cope with the challenge to safeguard their own brand identity when offering services on shared-platform models.[66] In order to compensate stagnating revenues from classic vehicle sales, automotive manufacturers need to foster their influence on the digital value-added chain. The in-house development of services and their integration into the vehicle are substantial cornerstones for becoming the lynchpin of the customer's mobile living environment. This most certainly applies when vehicle-related solutions can be expanded to other digital ecosystems such as the smart home or the working environment. The subsequent section shows the recent endeavors of OEMs and tier-n suppliers to approach digital transformation by investing in and partnering with established companies and digital rookies. Automotive players are seeking access to relevant technologies such as sensors, connectivity solutions, semiconductors, or artificial intelligence which depict the catalyzer for advancing ADAS and autonomous driving features. Since time-to-market is increasingly critical for automotive producers, organic growth and in-house setup of required capacities and

---

[65] cf. Cortada, James W. (2015) The Essential Manager S.40–43.
[66] cf. Gutzmer, Alexander (2018) Marken in der Smart City.

competencies is not always an option.[67] Given the rapid innovation cycles and dynamic development processes of software applications as well as the lacking trial-and-error mentality in complex corporate structures, leveraging new players is becoming essential to succeed, both in accelerated software development and AI. OEMs and suppliers have realized this opportunity and are forming cross-industry bundles (Fig. 5.9).

Continental's latest acquisition of ASC's high-resolution 3D Flash Lidar (Light Detection and Ranging system) business has expanded its portfolio of sensor technologies in form of laser-based distance measurement tools which enable automated driving features such as emergency brake assistants or collision warning.[68] Delphi also recognized the importance of lidar-based technologies needed for the mass-market

	Technologies			Enabling services	
Adaptive driver assistance systems	Infotainment	Human–machine interface	Communications, computing, and cloud	Connected vehicle services	Connected device services
OEMs (major automakers)					
Acquisition Audi/Daimler/BMW: Here (2015) GM: Cruise Automation (2016) *Investment* Volvo: Peloton (2016) *Partnership* Audi & Nvidia (since 2005) Bosch & TomTom (2015) GM & Mobileye (2015) VW & Mobileye (2015) BMW & Intel & Mobileye (2016) Hyundai & Cisco (2016)	*Investment* Ford: Livio (2013) *Partnership* Audi & Nvidia (since 2005)		*Partnership* Daimler & Qualcomm (2015) Hyundai & Cisco (2016) Toyota & KDDI (2016)	*Partnership* Ford & State Farm (2012) BMW & Pivotal (2015) Ford & Microsoft Azure (2015) Volvo & Microsoft (2015) Nissan & Microsoft Azure (2016)	*Acquisition* Daimler: Mytaxi (2014) GM: Sidecar (2016) *Investment* BMW: RideCell (2014) BMW: Zendrive (2014) GM: Telogis (2014) BAIC: Didi Chuxing (2016) Ford: Pivotal (2016) GM: Lyft (2016) Toyota: Uber (2016) VW: Gett (2016) *Partnership* BMW & Baidu (2015) BMW & Microsoft Azure (2016) Seat & Samsung & SAP (2016) Toyota & Microsoft Azure (2016)

Fig. 5.9 The supply side of connected cars: Deals, Investments, partnerships (part I). PwC' Strategy& (2016) S.35

---

[67] cf. UBS (2017) Longer Term Investments. Smart Mobility S.20ff.
[68] cf. PwC - Strategy& (2016) Connected car report 2016 S.37.

introduction of self-driving vehicles, thus conducting a strategic investment in Quanergy in 2015. In the same year, Valeo signed a technology cooperation agreement with Mobileye, an Israeli provider of driving assistance features,[69] to develop front-facing camera systems and sensor fusion that is capable of analyzing data drawn from multiple sensors to provide depth perception. Audi's venture into the non-automotive industry has already started in the mid-2000s, when the VW-affiliated OEM began partnering with Nvidia, back then known for their computer hardware, to speed up innovation cycles.[70] The cross-divisional team has developed a modular infotainment system decoupling software from hardware, eventually cutting down development time for new systems to 1 year (Fig. 5.10).

The success story of Harman, one of the industry's most profitable providers of infotainment systems and related services, is based on several critical strategic acquisitions—software companies S1nn and Symphony Teleca as well as Redbend in connected vehicle services—that have given access to over-the-air updates and cybersecurity technologies.

Among new entrants into the automotive supplier industry are also Cisco Systems and NXO Semiconductors, which partnered up in 2013 to invest in Cohda Wireless,[71] a company that specializes in wireless communication for automotive safety applications.

One of the most pivotal control points for the future profitability of digital business models is the human-machine interface. This technology layer through which the driver and passenger interact with the different onboard systems, e.g., the infotainment system, connectivity, and increasingly ADAS features, represents the virtual access to its users.[72] The primary focus of companies involves the effort to simplify the cockpit electronics by consolidating the cars electronic control units (Fig. 5.11).

However, OEMs and suppliers are actively integrating more specific HMI technologies from non-automotive companies such as Nuance,

---

[69] cf. Winkelhake, Uwe (2017) Die digitale Transformation der Automobilindustrie S.61–65.
[70] cf. Vo, Paul Hung (2015) Die Automobilindustrie und die Bedeutung innovativer Industrie 4.0 Technologien S.25ff.
[71] cf. PwC - Strategy& (2016) Connected car report 2016 S.37–39.
[72] cf. Skilton, Mark (2016) Building Digital Ecosystem Architectures S.51–53.

5 Business Model 2030: A Metamorphosis of the Automotive…

	Technologies				Enabling services	
Adaptive driver assistance systems	Infotainment	Human-machine interface	Communications, computing, and cloud	Connected vehicle services	Connected device services

New entrants from outside automotive

| Acquisition
Panasonic: Ficosa (2014)
Google: FCA (2016)
Nvidia: AdasWorks (2016)

New entrants
AdasWorks, Baselabs, Vector, Velodyne, Wind River | New entrants:
Apple, Baidu, Google | Investment
Intel: Omek (2013)

New entrants
Atmel, Fujitsu, Kyocera, LG, Toshiba | Acquisition
Cisco/NXP: Cohda Wireless (2013)

New entrants
Cohda Wireless, Kymeta, Veniam | Investment
Verizon: Hughes (2012)

Partnership
Airbiquity & Arynga (2016)

New entrants
Airbiquity, Allstate, Fleetmatics, Pivotal, Progressive, SiriusXM, Trimble, Verisk | Partnership
Daimler Moovel & IBM (2014)

Partnership
Airbiquity & Arynga (2016)

New entrants
Airbiquity, Apple, Contigo, Dash, Google, iTrack, Lyft, MyCarTracks, Uber |

Traditional suppliers

| Acquisition
Continental: Elektrobit (2015)
Delphi: Ottomatika (2015)
ZF: TRW (2015)
Continental: ASC (2016)

Investment
Delphi: Quanergy (2015)
Bosch: AdasWorks (2016)

Partnership
Valeo & Mobileye (2015) | Acquisition
Harman: Aha (2010)
Harman: S1nn (2014)
Continental: Elektrobit (2015)
Harman: Symphony Teleca (2015)

Partnership
Harman & Luxoft (2011)
Harman & Microsoft (2016) | Acquisition
Continental: Elektrobit (2015)

Partnership
Valeo & Safran (2013) | Acquisition
Bosch: ProSyst (2015)
Valeo: Peiker (2015) | Acquisition
Harman: Redbend SW (2015)
Harman: TowerSec (2016)

Partnership
Valeo & Capgemini (2015) | Acquisition
Harman: Aditi (2015)

(continued) |

Fig. 5.10 The supply side of connected cars: Deals, investments, partnerships (part II). PwC's Strategy& (2016) S.35/36

Immersion, or Myscript that offer a range of advanced features such as voice control, interactive touch, and haptics gestures as well as handwriting recognition to improve the user handling experience.[73]

Since companies, in general, not only engage in strategic partnerships to gain access to knowledge and technologies but to benefit from the international partner's supply and distribution channels, OEMs and suppliers also have to face limitations to the free, globalized market. China

---

[73] cf. Herrmann, Andreas; Brenner, Walter; Stadler, Rupert (2018) Autonomous Driving S.62ff.

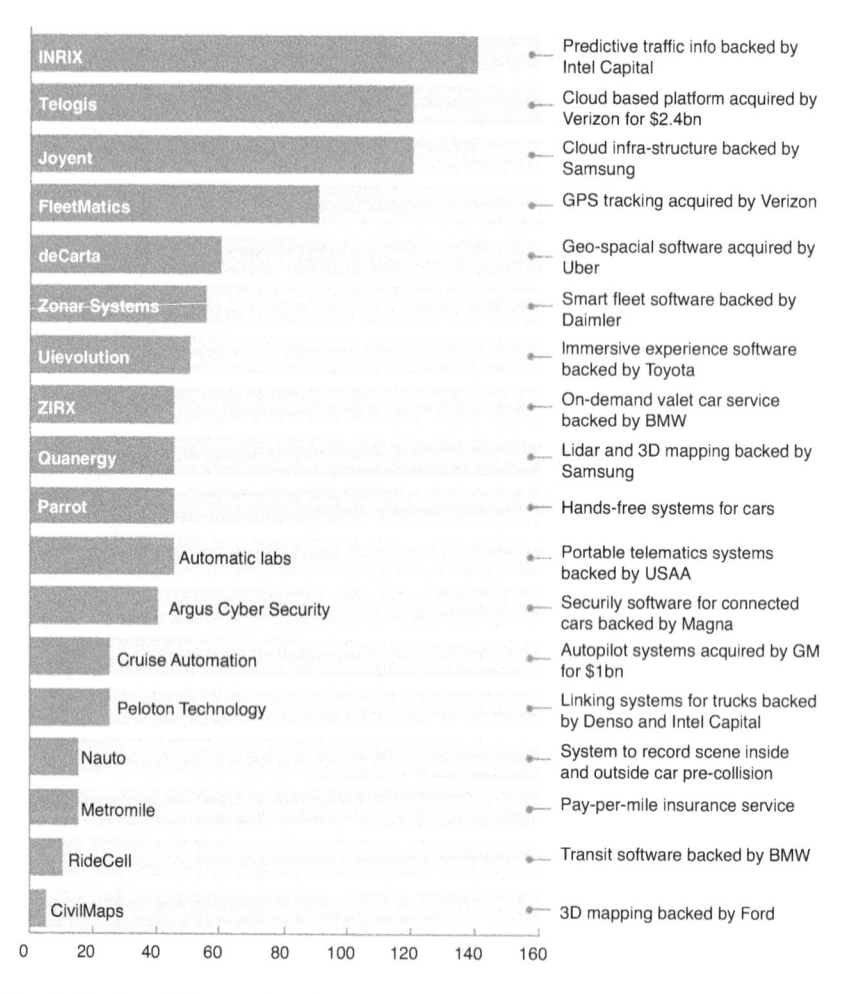

Fig. 5.11 Acquisitions of and investments in autonomous driving capabilities. McKinsey & Company I (2017) S.20

represents a critical market to the future growth strategies of every automotive-related company.[74] However, the Chinese government has erected barriers to foreign participation in form of additional taxes or restrictions

---

[74] cf. PwC - Strategy& (2016) Connected car report 2016 S.13/25.

and regulations to cross-market acquisition while offering considerable support to the domestic industry by providing testing facilities or subsidies. The protectionist measures also require that at least 50% of a newly established joint venture between domestic and foreign companies has to stay in local control. Recent announcements of the Chinese communist party, in fact, have confirmed an iterative cutback on protectionist regulations until 2022, but international OEMs and suppliers are already in the process of gaining access to the market.[75] Telecommunication giant Huawei technologies has signed agreements with Dongfeng and Chongqing Changan, both automobile producers, to establish cooperation on vehicle connectivity and autonomous driving. Due to China's censorship of Google and its affiliated mapping application, Audi started partnering with Tencent, which operates the country's widely popular messaging service WeChat to allow location sharing in vehicles. The PSA group, best known for its brands Peugeot and Citroen, will equip its models with Wi-Fi hotspots in collaboration with Alibaba while China Mobile and the German Telekom are co-developing a platform for internet-connected cars.

Traditional business models that are limited to developing, building, and retailing cars have widely become obsolete. Niche manufacturers might still be successful in pursuing a special value proposition, but mass market OEMs need to advance in newly formed fields such as fleet operations or digital services, or retail mobility.[76] The extent of their restructuring requirements will depend on their overall value chain integration. Affiliated automotive groups face an upheaval of their corporate structure. The degree to which these companies manage to adopt their value creation process including all value chain functions and are capable of building digital know-how will be distinguishing factors in the survival of the fittest (Fig. 5.12).[77]

---

[75] cf. Deloitte I (2017) The Future of the Automotive Value Chain S.27–33.
[76] cf. Iskander Business Partner (2016) Digitalisierung in der Automobilindustrie S.6/24–25.
[77] cf. Deloitte I (2017) The Future of the Automotive Value Chain S.13/18ff.

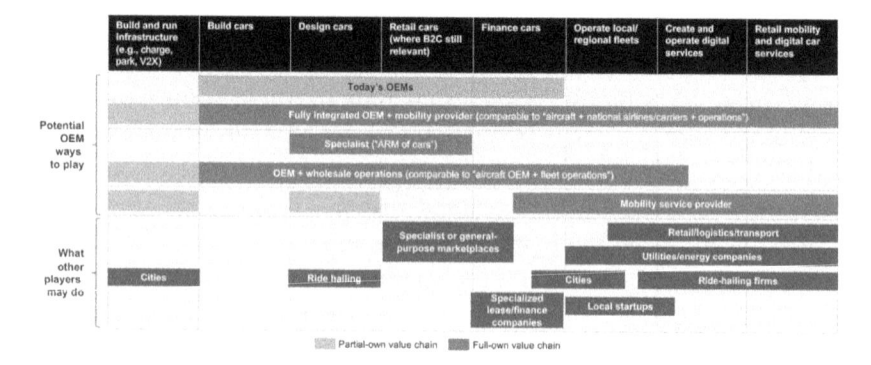

Fig. 5.12 Different models of value chain integration. PwC's Strategy& (2017) S.20

5.3 Defining Use Cases in the Disrupted Automotive Market

5.3.1 Determining the Accessibility of new Revenue Sources

The graphic on the next page depicts the allocation of yearly expenditures of an average Western higher-income household, gradually highlighting spending directly or indirectly associated with private mobility. Three categories are accentuated, clustered into digital service opportunities and assigned a forecasted expenditure value in 2030. As indicated in the first category, gross addressable expenditures from an augmentation of vehicles through the addition of digital services directly associated with the use of the car are predicted to decline by approximately 18.2%. Providing functions as a service to increase customer loyalty and achieve higher customer lifetime value will become a commodity, that is to say, additional functionality is no measure of differentiation but seen as a "must-have." Customers can easily switch brands in the highly competitive market, hence, OEMs need to offer updated functions with a low mark-up. Cross-modal mobility service, as presented in the scenario and smart city analyses primarily include on-demand hailing and car sharing

5 Business Model 2030: A Metamorphosis of the Automotive... 125

services.[78] The main factor leading to a stagnation or respectively a tendency to a slight decline in accumulated household spending for this category is a price erosion. Multiple new entrants have already accessed the shared mobility market resulting in a price competition. Besides low profit margins, car sharing providers need to reach a critical scale and a certain market share to achieve sufficient Return on Investments (ROI).[79] According to the depicted forecast, expenditures for fifth-screen ecosystem services are likely to increase by roughly 27.5%, thus representing a highly attractive business volume for OEMs and third parties. Providing and exerting power over HMIs and vehicle-integrated control points to market own services and offer a platform for commercial transactions is the key driver for successful and profitable market positioning (Fig. 5.13).

The new emphasis on service revenues implies the OEMs' need to adjust to newly allocated revenue streams over the product lifecycle. Today, cashflows are primarily generated by the traditional sales of cars and spare parts in the aftersales area.[80] In the medium term, continuous cash flows will shift progressively towards in-car services, in-app sales, and mobility service solutions.[81] Pay-as-you-drive payment methods and subscription-based models for streaming services or traffic service features

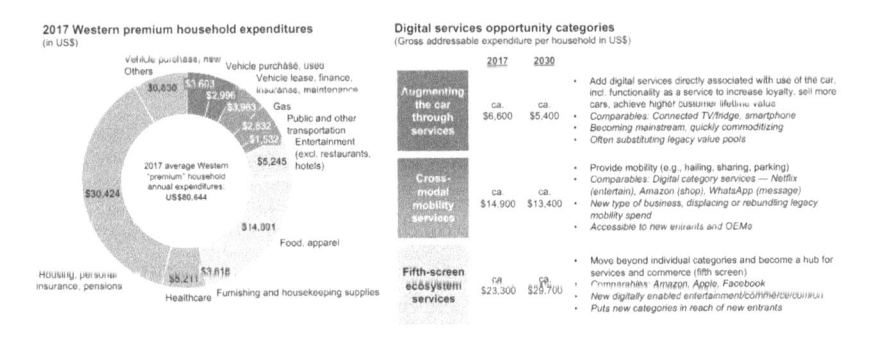

Fig. 5.13 Roboconomy digital service opportunities. PwC's Strategy& (2017) S.23

---

[78] cf. Roland Berger (2014) Autonomous Driving.
[79] cf. KPMG I (2018) Autonomous Vehicle Readiness Index S.24ff.
[80] cf. Roland Berger (2015) Maßkonfektion im Aftersales S.13–15.
[81] cf. A.T. Kearney (2016) How Automakers can survive the Self-Driving Era S.6–9.

will lead to punctiform revenues that occur at the actual time of usage. The access to certain revenue pools, which will be illustrated in the next sections, depends on several prerequisites regarding the usage and ownership of data streams. Stakeholders such OEMs, insurance companies, law enforcement agencies, financial institutions, or digital start-ups are eager to gain access to the metadata generated that is crucial to develop,[82] manage and improve new business models in the automotive landscape. Traditionally, customer transaction data is controlled by access network operators while the data inflow of application and service providers is limited to data on user profiles. A crucial challenge for automotive manufacturers will be to occupy the relevant customer interfaces to comprehensively gather exploitable data and apply big data analytic tools.[83] Another concern, that is indirectly linked to data ownership is the definition of Intellectual Property (IP) which helps determine industry standards and protocols to set a base for interoperable systems and devices. Legislative institutions need to establish a legal foundation, specifically tailored for the new standards in the data-driven automotive business environment, that clearly settle legal claims and provide transparent terms and conditions. Furthermore, distinct data and personal rights could reinforce the customers' trust in data-based business models since their willingness to divulge data will become a primary means for many companies to generate revenues.[84] Privacy and cybersecurity issues, as indicated in Chap. 4, have significant value for most consumers and, thus, need to be addressed on a political and legislative level (Fig. 5.14).

In preparation for appraising medium and long-term revenue pools, the diagram above illustrates a simplified view of the current areas of operations in which OEMs, automotive supplier, tech companies, and specialist service firms are competing increasingly. While car manufacturers were traditionally focused on the development and sales of hardware with a new emphasis on advanced automation features, connectivity packages, and tools such as preventive maintenance diagnosis,[85] novel

---

[82] cf. EY (2014) Deploying autonomous vehicles S.6–7.
[83] cf. Linnhoff-Popien, Claudia; Schneider, Ralf (2018) Digital Marketplace unleashed S.633ff.
[84] cf. Anderson, James et al. (2016) Autonomous Vehicle Technology S.94–97.
[85] cf. Huber, Walter (2016) Industrie 4.0 in der Automobilproduktion S.118ff.

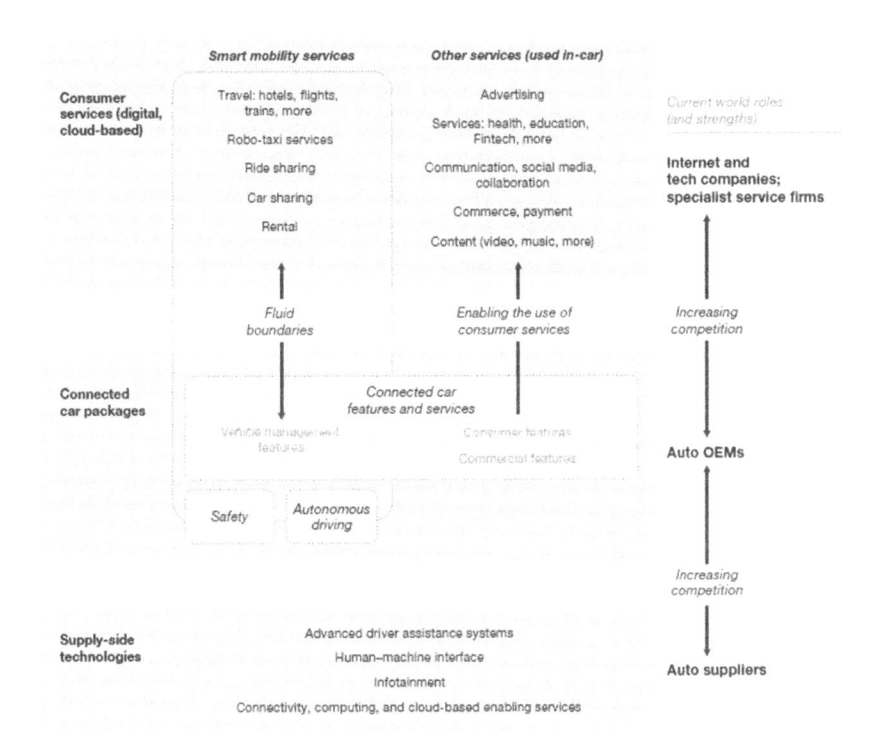

5 Business Model 2030: A Metamorphosis of the Automotive… 127

Fig. 5.14 The full range of connected car technologies and services. PwC's Strategy& (2016) S.15

forms of digital, cloud-based consumer services have emerged that result in an increasingly heterogenous competition in the automotive business landscape. New market entrants are rivaling with incumbent players about nascent sources of revenues and shares in overall profit generation. Connected car features, especially the HMI, drive the use of intersectoral consumer service.[86] HMIs enable targeted advertisements based on the vehicle location, destinations, and driving patterns, social media providers are coupling their applications with on-board screens and content providers are offering music and video streaming services.[87] OEMs do

---

[86] cf. PwC (2018) Five trends transforming the Automotive Industry S.23–27.
[87] cf. Siegfried, Patrick (2014) Analysis of the service research studies in the German research field S.94–104.

not seek to intervene in unknown industries and provide own, in-house developed services that require a diverging set of core competencies, but rather try to access revenue streams that emerge from the usage of the vehicle-integrated system and human–machine interfaces. Control points within the car, such as touch screens and on-board devices are the essential medium to market those services.

The other main business segment which is coined by more fluid boundaries between traditional OEMs and new competitors is smart mobility services. OEMs have recognized the consumer interest in on-demand and shared mobility solutions and are developing own car-sharing models.[88] Despite the fact that new mobility services will indeed capture new revenue streams, a certain degree of cannibalization of vehicles sales that would have been conducted with private persons will be inevitable. Nevertheless, OEMs have already launched branded carpooling and -sharing services such as DriveNow (BMW), car2go (Mercedes-Benz/Smart) or MultiCity (Citroen), personalized micro-car-sharing options (Audi Unite) or personalized dynamic carpooling (Audi Select). However, third parties are operating cross-brand car-sharing services or platforms that automatically pool customers with drivers such as CarJump, CiteeCar, Flinkster, book-n-drive, Uber, or clever-shuttle.[89] The subsequent subchapter will assess the profitability and growth potential of traditional and novel digital business models and indicate the prospective shift of the current revenue structure of automotive manufacturers.[90]

5.3.2 Redistribution and Monetization of Revenue Pools

The illustration below gives an overview of the clustered revenue structure in the automotive industry that will be assessed in more detail. Emanating from a high disruption scenario, the automotive revenue pool

---

[88] cf. McKinsey & Company (2014) Connected car, automotive value chain unbound S.38–40.
[89] cf. McKinsey & Company II (2015) Urban mobility at a tipping point S.15–18.
[90] cf. Diehlmann, Jens; Häcker, Dr. Joachim (2013) Automotive Management S.218ff.

will likely grow and diversify significantly. The three revenue segments combined could add up to roughly USD 6.7 trillion, provided a yearly growth rate of 4.4% (Fig. 5.15).

In this calculation, one-time vehicle sales will indeed decline on a global level despite the increasing vehicle unit sales driven by macroeconomic growth in emerging economies, but revenues generated by price premiums for alternative powertrains and autonomous driving technology features will clearly exceed the amount necessary to compensate for the marginal regression in overall one-time vehicle sales quantity.[91]

Given a scenario in which shared mobility has spread in urban regions, the higher capacity utilization of shared vehicles leads to increased annual maintenance spending, expanding the total turnover in the aftersales market.[92]

In spite of lower maintenance requests for electric powertrains and regressive expenditures on repair work for autonomous vehicles due to their technological maturity and growing market penetration, the overall revenue pool is expected to reach up to USD 1.2 trillion. The forecast suggests that up to 30% of the total volume of business in 2030 could be generated by recurring revenues resulting from pay-as-you-drive and subscription-based models of shared mobility services and digital service applications.[93] Since 2030 is still more than a decade away, scenario evaluations have a very hypothetical character. The subsequent financial evaluation will therefore be based on a 2020 base case scenario (Fig. 5.16).

Regarding connectivity hardware, OEMs are challenged by the consumer electronics industry that is seeking to transfer their products onto vehicle systems. The overall life-cycle bucket can be clustered into five distinct areas of connectivity features that comprise navigation systems, smartphone integration and application processes, entertainment, remote services and advanced driving assistance systems. The depicted quantitative forecasts consider an increasing adoption rate of navigation systems but account for price erosions for today's built-in feature sets leading to a

---

[91] cf. McKinsey & Company III (2016) Monetizing car data S.11/16/23.
[92] cf. McKinsey & Company II (2017) The changing aftermarket game.
[93] cf. Iskander Business Partner (2016) Digitalisierung in der Automobilindustrie S.6/24–26.

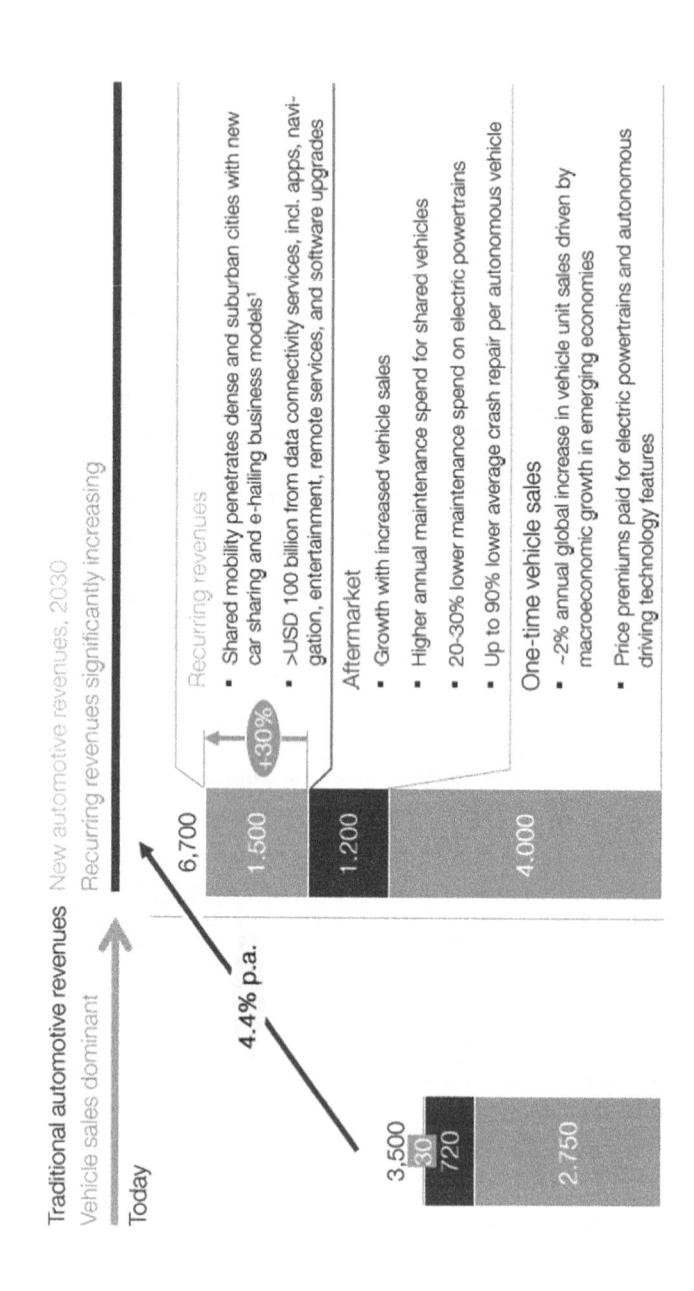

Fig. 5.15 The growth potential of the automotive revenue pool (high-disruption scenario). McKinsey & Company I (2016) S.6

5 Business Model 2030: A Metamorphosis of the Automotive... 131

Representative D-segment premium vehicle, Germany   BASE CASE

	Navigation	Smartphone integration/ app access	Entertainment	Remote services	ADAS	Total
Connectivity hardware revenues 2020[1] EUR	~1,440	~560	~200	~440	~830	~3,470
Description/ assumptions 2020	• Built-in navigation system • Real-time traffic info and prediction • Advanced car integration	• In-car app access • Full smartphone integration	• Large screens and integrated user interface • Music/video streaming • Access to home media	• Restricted car usage • Car steering via mobile, e.g., parking	• Adaptive cruise control • Emergency breaking • Active lane assist	

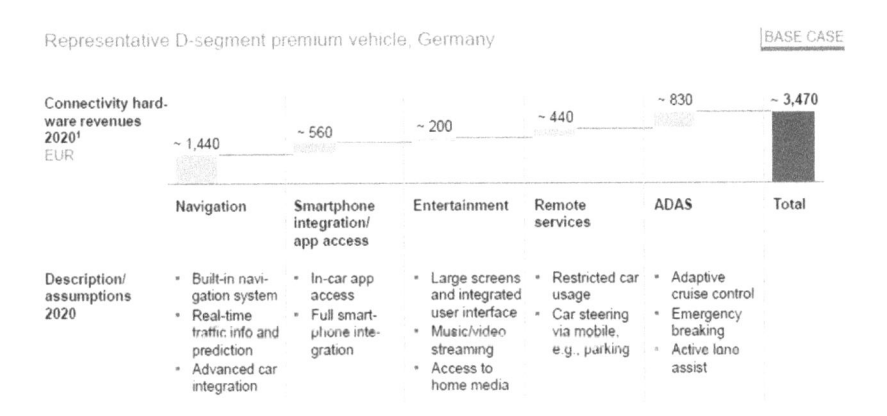

Fig. 5.16 Growth of upfront connectivity hardware revenues. McKinsey & Company (2014) S.26

stagnation in revenues.[94] As navigation becomes standard equipment, OEM will be unable to monetize branded systems unless they continue to improve the integration of features into the overall user experience along with superior interface hardware which facilitates the commercialization on a premium price level. The seamless integration of smartphones and further portable devices combined with full in-car access to applications, both installed on user devices as well as directly on the vehicle's operating system, represents another major revenue bucket.[95] However, third-party offerings will trigger a strong proliferation of solutions in the future resulting in potentially drastic price drops, especially if these digital innovators decide to provide their platform free of charge. AppleCarPlay or MirrorLink are already competing intensely with automotive manufacturers regarding pricing tactics and mark-ups (Fig. 5.17).

If automatic over-the-air updates are established in mass market vehicles and OEMs' shares in revenues are not cannibalized by navigation applications provided by digital natives, fees from map updates or commissions for location-based recommendations could offset today's revenues from pre-installed offline maps. As every smartphone and other mobile end devices are preloaded with free maps applications

---
[94] cf. Herrmann, Andreas; Brenner, Walter; Stadler, Rupert (2018) Autonomous Driving S.138ff.
[95] cf. Bundesverband Digitale Wirtschaft (2016) Connected Cars – Geschäftsmodelle S.13ff.

Representative D-segment premium vehicle, Germany | BASE CASE

"Driver's time and attention" revenues per vehicle, 2020¹ EUR

Navigation ~270 | Apps ~70 | Entertainment ~140 | Total ~480

| Description/assumptions | • Commissions for location-based recommendations
• Fees from map updates
• Monetization depends on success of built-in navigation systems | • Users pay for in-app content, e.g., subscriptions or virtual goods
• OEMs unable to monetize data from third-party apps, similar to telco providers | • Shared audio/video streaming fees
• Streaming content provided by third parties
• OEMs have control over HMI integration | |

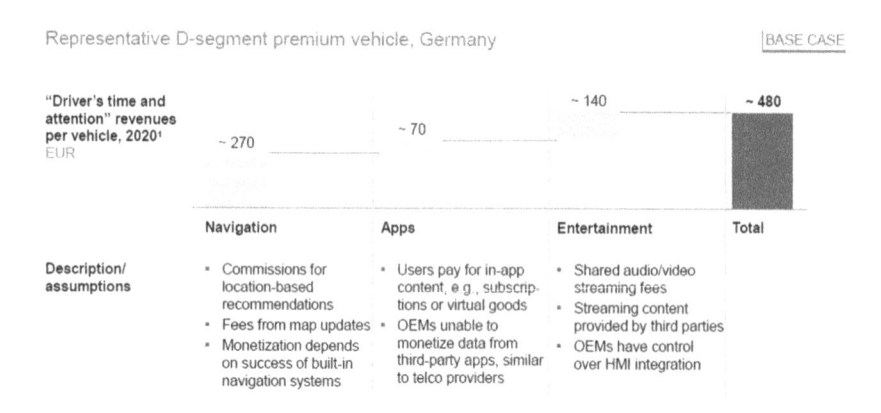

Fig. 5.17 Future usage-based revenue potential. McKinsey & Company (2014) S.27

that offer advanced navigation features such as crowd-based traffic information and routing, paid navigation systems will likely not be able to continue their commercial success.[96] In order to engage in revenue streams emanating from application ecosystems, OEMs need to exert a certain degree of control over the HMI integration.[97] Additional value generated by in-app contents through subscriptions or virtual goods is typically shared by the developer and the platform operator. Automotive companies should capture the role of an intermediate to access this revenue pool (Fig. 5.18).

Research suggests that 23% of new-car buyers are willing to follow maintenance app recommendations provided by the respective automotive brand, which equates to up to 760 Euro of revenue redistribution in the overall vehicle life-cycle. Beyond that, car condition data represents an opportunity to increase customer loyalty and optimize product and durability specifications[98] throughout the development processes.

As indicated in previous chapters, OEMs' set-up of a sustainable revenue access requires the occupancy of certain vehicle-internal and external control points. These will provide different levels of perceived

---

[96] cf. A.T. Kearney (2016) How Automakers can survive the Self-Driving Era S.6–10.
[97] cf. Capgemini Consulting (2017) Beyond the Car S.12ff.
[98] cf. Brynjolfsson, Erik; Mcafee, Andrew (2014) The Second Machine Age S.92–97.

5 Business Model 2030: A Metamorphosis of the Automotive... 133

EUR for representative D-segment premium vehicle, Germany   BASE CASE

Average maintenance spending over 5-year car life cycle[1]

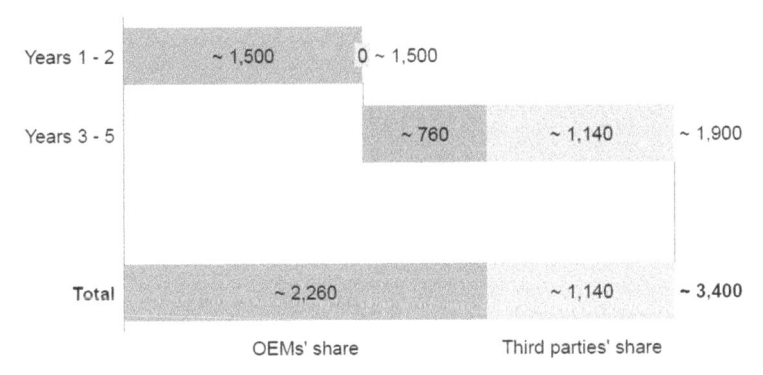

Years 1 - 2	~1,500	0 ~1,500	
Years 3 - 5		~760 ~1,140	~1,900
Total	~2,260	~1,140	~3,400
	OEMs' share	Third parties' share	

Fig. 5.18 Potential of post-warranty revenue re-distribution. McKinsey & Company (2014), S.28

customer value depending on the brand positioning, target market, and current business scale of the automotive manufacturer (Fig. 5.19).[99]

The figure above sub-divides potential control points of various areas into two strategic groups, "must-haves" and "differentiators." Must-have commodities inside the vehicle such as content applications, CPUs, or control units as well as cloud-related features including mobile connection, data gateway, and granular map data require high investments for proprietary solutions and provide limited differentiation potential. IT companies and digital natives are more suited to develop adequate software solutions based on their core competencies and economies of scale in developing digital features. In this respect, sourcing agreements and partnerships between software providers and OEMs ensure an integrated and competitive provision of must-have features to the customer.[100] The occupation of data gateway control points, in particular, offers the opportunity to position the brand as a trustful data hub.[101] Customer concerns about data usage and security are significant hurdles to overcome to

---

[99] cf. McKinsey & Company (2014) Connected car, automotive value chain unbound S.21–27.
[100] cf. Attias, Danielle (2017) The Automobile Revolution S.69–72.
[101] cf. BearingPoint Institute I (2017) Create a "Connection" with your Customer S.5.

Area	Potential control points	Strategic assessment	Rationale
Inside the car	HMI	Differentiator	• Key for customer experience and branding
	Apps/content	"Must-have" features (via sourcing/partnering)	• High investments for proprietary solution • Difficult to compete against IT giants • Difficult to reach economies of scale
	App store		
	IVI[1] OS		
	CPU/control unit		
	Car position data		
	Car sensor data	Differentiator	• OEM core competency • Key for ADAS features
	Car actuators		
Connection to the cloud	Mobile connection	"Must-have" features (via sourcing/partnering)	• High investments for proprietary solution • Limited differentiation potential for "commodity" features • Limited competency
Data cloud	Data gateway		
	Granular map data		
	Dynamic, real-time geospatial information	Differentiator	• Key for ADAS features • Potential competitive advantage

Fig. 5.19 Control points and decisive key differentiators. McKinsey & Company (2014), S.33

achieve customer acceptance of all data-driven digital business models.[102] OEMs could be able to generate a competitive advantage by conducting data handling processes in a transparent and traceable way that provides for a unique selling proposition as a trusted brand in contrast to many well-known IT giants that regularly suffer from customer data abuse scandals. A full disclosure of the individual vehicle-related data usage could thus reinforce brand and customer loyalty. In conclusion, there are five ways to monetize connected services in addition to revenues from traditional vehicle sales. Car-integrated connected packages on the one hand directly contribute to the business volume but on the other hand also provide data that can be used to increase internal efficiency, quality, and product differentiation. This, in turn, reinforces customer loyalty and facilitates the defense of list price levels and premium mark-ups. Furthermore, in-house created databases of customer information can be monetized through future business models, especially in mobility services

---

[102] cf. Goodyear; London School of Economics I (2015) Autonomous vehicles: Negotiating a place on the Road S.4ff.

and multi-modal transportation options.[103] The last step of implementing a platform-based business model is the inclusion of partners into a comprehensive ecosystem of consumer services that will enable access to shared service revenues.[104]

5.3.3 Excursus: Upheaval in the Insurance Industry

The positioning of advanced connectivity and ADAS features in the medium term and autonomous driving in the long term will cause a decisive upheaval of the car insurance industry. The calculation of automobile liability insurance based on a flat-rate model which is today's standard practice will migrate towards a pay-as-you-drive model.[105] Actually, this model comprises two different modes of calculation, a tariff classification determined by the risk evaluation of the driving behavior and secondly, a usage-based charge on a time or distance base. Pay-how-you-drive and pay-when-you-drive models thus offer an incentive structure for a safe manner of driving[106] that is already offered to novice drivers in several European countries and is highly suitable for on-demand shared mobility services.

As vehicles are incrementally more connected and are capable of operating various driving situations highly automated or autonomously, the traditional service provision between insurances and private persons will shift towards a new interface between insurances and OEMs. Accidents or traffic incidents that are not driver-inflicted but rather caused by technical defects or vehicle system failure will be assigned to the OEM's scope of liability (Fig. 5.20).[107]

But both parties can derive mutual benefit from this new business relationship. Automotive manufacturers can leverage their data provision position to help insurances reduce costs and generate additional revenues,

---

[103] cf. Meyer, Gereon; Shaheen, Susan (2017) Disrupting Mobility S.149–151.
[104] cf. IBM Center for Applied Insights (2015) Digital disruption and the future of the automotive industry S.13–16.
[105] cf. EY (2017) The evolution in self-driving vehicles S.5.
[106] cf. McKinsey & Company III (2016) Monetizing car data S.25.
[107] cf. Anderson, James et al. (2016) Autonomous Vehicle Technology S.118.

Reducing costs	Generating revenues		
Driving behaviors Improve the risk profile Offer incentives for the autonomous driving portion of a customer's driving behavior (based on fewer human driving errors)	**Client segmentation** Expand into new areas; be a first mover in autonomous driving by offering a unique selling proposition	**Innovative value-added services** Add new services close to the core business of insurance	**Customer loyalty** Reduce customer turnover with additional services and tailored premium offers

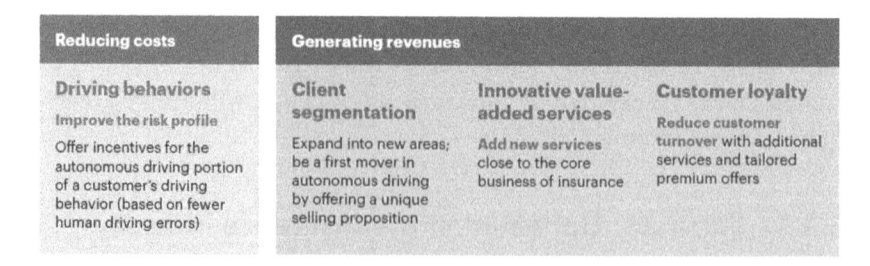

Fig. 5.20 Four ways insurers can use data to improve profitability. A.T. Kearney (2016), S.29

Traditional risk-pricing model		Potential new pricing model	
• Age	• Declared distance	• Number of driving hours	• Maintenance
• Sex	• Garage	• Time	• Parking
• Driving years	• Claims	• Distance and location	• Weather conditions
• Declared usage		• Velocity or limit control	
Categorizing customers in a few segments based on static parameters such as sex, age, and location due to the limited availability of data		Categorizing customers in detailed segments based on new parameters such as speed and sprinting	

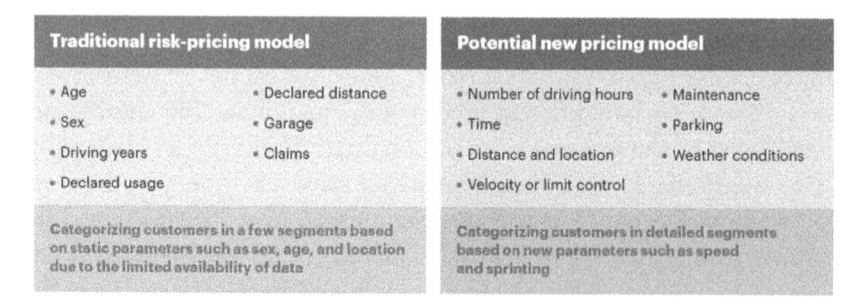

Fig. 5.21 Traditional risk-pricing vs. usage-based pricing models. A.T. Kearney (2016) S.35

thus increasing overall profitability. Data on driving behavior, stored and analyzed by system-integrated hardware and software applications, facilitates a more sophisticated customer risk management and the creation of more distinct customer risk profiles.[108] New forms of revenue generation are based on a more differentiated client segmentation, innovative value-added services as well as an increase in customer loyalty due to data-based individual service offerings (Fig. 5.21).

For insurances that provide usage-based policy programs, access to comprehensive data is crucial to segment and pool different types of drivers effectively. Traditional risk-pricing models categorizing clients into standardized segments based on static parameters such as age, sex, driving experience, or average distances per year are not suitable to calculate

---

[108] PA Consulting (2017) Autonomous Driving S.11–13.

personalized policy fees. Potential new pricing models could incorporate parameters that are accessible through data processing tools such as number of driving hours, traveled distances, velocity and speed limit control, or weather conditions.[109] Further promising spheres of activity are predictive maintenance and settlement of claims.

Automated emergency calls that have become compulsory as a regulatory driver for the diffusion of connectivity technology and comprehensive sensor systems that record accidents in real-time, facilitate a faster and improved provision of medical care. Beyond that, sensory data can be stored and analyzed in form of crash protocols which allows insurances to compile a list of damaged vehicle components and calculate the resulting repair outlays. This automated calculation of accidental damage leads to a substantial time and cost reduction in claim adjustment. Predictive maintenance features refer to an early warning system for deterioration or abrasion of vehicle components[110] whose conditions can be traced by integrated sensors. Customers are notified in case of pending maintenance operations, early enough to avoid breakdowns or consequential damage. Thus, big data applications lead to an increase in customer loyalty advanced value-added services.[111]

Already today, several insurance providers that offer usage- and telematic-based policies are competing in the German market. Allianz Bonus Drive, AXA Drive Check, AdmiralDirekt Telematic option, or Mercedes-Benz Bank pay-how-you-drive have seized a chance to address new forms of shared mobility insurance models and offer a differentiating value proposition for certain customer segments. But OEMs and traditional insurance companies have yet to overcome three main hurdles. Since empirical values are limited and benchmark values have not been established yet, a risk-based calculation of tariffs and the correct interpretation of gathered driving data pose a cumbersome challenge. In the long term, regulators need to conceptualize a new legislation that covers all liability issues for OEMs and insurances. This complex and time-consuming task will likely delay the marketability of novel insurance

---

[109] cf. McKinsey & Company III (2016) Monetizing car data S.25–27.
[110] cf. Linnhoff-Popien, Claudia; Schneider, Ralf (2018) Digital Marketplace unleashed S.297.
[111] cf. PwC (2018) Five trends transforming the Automotive Industry S.23.

policies and hence, the introduction of autonomous driving features. The third challenge to overcome is customer concerns regarding an intrusion of privacy and their willingness to share personal data. As addressed in Chap. 4, certain customer groups are likely to deny access to relevant usage data, thus impeding advanced insurance services. OEMs seeking to make money out of the transfer of vehicle-induced data sets to insurance companies have to dispel consumers' doubts in order to generate sustainable revenue streams.

References

Anderson, J., Kalra, N., Stanley, K., Sorenson, P., Samaras, C., & Oluwatola, O. (2016). *Autonomous vehicle technology. A guide for policymakers.* Rand Corporation.

Attias, D. (2017). *The automobile revolution. Towards a new electro-mobility paradigm.* Springer Int. Publishing AG.

BearingPoint Institute I. (2017). *Create a "connection" with your customer. Why retailers must adapt to changing consumer demands*, S. 1–12. Retrieved March 16, 2021, from https://www.bearingpoint.com/files/007_21_BNR_Report-1_Create-a-connection-with-your-customer.pdf&download=0&itemId=495640

BearingPoint Institute II. (2017). *Re-thinking the European business model portfolio for the digital age*, S. 1–16. Retrieved March 16, 2021, from https://www.bearingpoint.com/files/009_01_DEM_Re-thinking_the_European_Business_Model_Portfolio_for_the_Digital_Age.pdf&download=0&itemId=495958

BearingPoint, IIHD Institut für internationales Handels- und Distributionsmanagement. (2017). Ecosysteme & Plattformen verändern die Handelslandschaft. Wie branchenübergreifende Koopertionen die Wettbewerbsstrukturen und -logiken des Handels von morgen bestimmen. In *Retail & Consumer* (No.12), S. 1–24. Retrieved April 18, 2021, from https://www.bearingpoint.com/files/BEDE17_1148_RP_DE_Ecosysteme_und_Plattformen_ver%C3%A4ndern_die_Handelslandschaft.pdf&download=1&itemId=461481

Brynjolfsson, E., & Mcafee, A. (2014). *The second machine age. Wie die nächste digitale Revolution unser aller Leben verändern wird.* Plassen Verlag.

Bundesverband Digitale Wirtschaft. (2016). *Connected Cars—Geschäftsmodelle*, S. 2–15. Retrieved May 11, 2021, from https://www.bvdw.org/fileadmin/bvdw/upload/publikationen/digitale_transformation/Diskussionspapier_Connected_Cars_Geschaeftsmodelle.pdf

Capgemini Consulting. (2017). *Beyond the car*, S. 1–36. Retrieved February 25, 2021, from https://www.capgemini.com/consulting-de/wp-content/uploads/sites/32/2017/05/cars-online-study-2017.pdf

Caudron, J., & Peteghem, D. (2018). *Digital transformation. A model to master digital disruption* (3. Aufl.) N.A.

Cordon, C., Garcia-Mila, P., Vilarino, T., & Caballero, P. (2016). *Strategy is digital. How companies can use big data in the value chain*. Springer International Publishing AG.

Cortada, J. W. (2015). *The essential manager. How to thrive in the global information jungle*. John Wiley & Sons.

Deloitte I. (2017). *The future of the automotive value chain. 2025 and beyond*, S. 1–64. Retrieved April 4, 2021, from https://www2.deloitte.com/content/dam/Deloitte/us/Documents/consumer-business/us-auto-the-future-of-the-automotive-value-chain.pdf

Deloitte University Press. (2015). *Patterns of Disruption. Anticipating disruptive strategies in a world of unicorns, black swans and exponentials*, S. 1–34. Retrieved May 11, 2021, from https://www2.deloitte.com/content/dam/Deloitte/br/Documents/technology/Patterns-of-disruption.pdf

Deutsche Bank AG. (2017). The digital car. More revenue, more competition, more cooperation. In *German Monitor The digital economy and structural transformation*, S. 1–36. Retrieved March 16, 2021, from https://www.dbresearch.com/PROD/RPS_EN-PROD/PROD0000000000446248/The_digital_car%3A_More_revenue%2C_more_competition%2C_m.pdf

Diehlmann, J., & Häcker, J. Dr. (2013). *Automotive management. Navigating the next decade* (2. Aufl.). Oldenbourg Verlag.

Diez, W. (2018). *Wohin steuert die deutsche Automobilindustrie?* (2. Aufl.). Walter de Gruyter GmbH.

Ecola, L., Zmud, J., Gu, K., Phleps, P., & Feige, I. (2015). *The future of mobility. Scenarios for China in 2030*. Rand Corporation.

EY. (2014). *Deploying autonomous vehicles. Commercial considerations and urban mobility scenarios*, S. 1–8. Retrieved March 3, 2021, from https://www.ey.com/Publication/vwLUAssets/EY-Deploying-autonomous-vehicles-30May14/$FILE/EY-Deploying-autonomous-vehicles-30May14.pdf

EY. (2017). *The evolution in self-driving vehicles. Trends and implications for the insurance industry*, S. 1–12. Retrieved April 18, 2021, from https://www.ey.

com/Publication/vwLUAssets/ey-self-driving-vehicle-v2/$FILE/ey-self-driving-vehicle-v2.pdf

Fisher, T. (2009). *The data asset. How smart companies govern their data for business success.* John Wiley & Sons, Inc.

Gerdes, C., & Maurer, M. (2015). *Autonomous driving. Technical, legal and social aspects.* Springer Verlag GmbH.

Goodyear, London School of Economics I. (2015). Autonomous vehicles: Negotiating a place on the Road. Survey Info Graphic. In *Think good mobility*, S. 1–4. Retrieved February 2, 2021, from http://docs.wixstatic.com/ugd/efc875_2fd46c7db57640c184e89a7e8778fcb2.pdf

Gutzmer, A. (2018). *Marken in der Smart City. Wie die Cyber-Urbanisierung das Marketing verändert.* Springer Fachmedien GmbH.

Hassan, Q. (2018). *Internet of things—A to Z. Technologies and applications.* John Wiley & Sons, Inc.

Herrmann, A., Brenner, W., & Stadler, R. (2018). *Autonomous driving. How will the driverless revolution change the world.* Emerald Publishing Ltd.

Huber, W. (2016). *Industrie 4.0 in der Automobilproduktion: Ein Praxisbuch.* Springer Fachmedien GmbH.

IBM Center for Applied Insights. (2015). *Digital disruption and the future of the automotive industry*, S. 1–16. Retrieved June 2, 2021, from https://www-935.ibm.com/services/multimedia/IBMCAI-Digital-disruption-in-automotive.pdf

Iskander Business Partner. (2016). *Digitalisierung in der Automobilindustrie. Wer gewinnt das Rennen? Traditioneller Automobilhersteller oder Silicon Valley?* S. 3–28. Retrieved July 3, 2021, from http://i-b-partner.com/wp-content/uploads/2016/08/2016-09-06-Iskander-RZ-Whitepaper-Digitalisierung-in-der-Automobilindustrie-DIGITAL.pdf

Jaekel, M. (2015). *Smart City wird Realität. Wegweiser für neue Urbanitäten in der Digitalmoderne.* Springer Fachmedien GmbH.

Karls, I., & Mueck, M. (2018). *Networking vehicles to everything. Evolving automotive solutions.* Walter de Gruyter GmbH.

Kearney, A. T. (2016). *How Automakers can survive the Self-Driving Era. A.T. Kearney study reveals new insights on who will take the pole position in the $560 billion autonomous driving race*, S. 1–36. Retrieved March 16, 2021, from https://www.kearney.com/automotive/article?/a/how-automakers-can-survive-the-self-driving-era

KPMG I. (2016). *I see. I think. I drive. (I learn). How Deep Learning is revolutionizing the way we interact with our cars*, S. 1–44. Retrieved March 16,

2021, from https://assets.kpmg.com/content/dam/kpmg/se/pdf/komm/2016/se-isee-ithink-idrive-ilearn.pdf

KPMG I. (2017). *Reimagine places: Mobility as a Service. The Mobility as a Servce (MaaS) Requirements Index 2017*, S. 1–32. Retrieved July 2, 2021, from https://assets.kpmg.com/content/dam/kpmg/uk/pdf/2017/08/reimagine_places_maas.pdf

KPMG III. (2017). *Islands of Autonomy. How autonomous verhicles will emerge in cities around the world*, S. 1–28. Retrieved July 22, 2021, from https://assets.kpmg.com/content/dam/kpmg/za/pdf/2017/11/islands-of-autonomy-web.pdf

KPMG I. (2018). *Autonomous vehicle readiness index. Assessing countries' openness and preparedness for autonomous vehicles*, S. 1–60. Retrieved July 22, 2021, from https://assets.kpmg.com/content/dam/kpmg/nl/pdf/2018/sector/automotive/autonomous-vehicles-readiness-index.pdf

Kreutzer, R., Neugebauer, T., & Pattloch, A. (2017). *Digital business leadership. Digital transformation, business model innovation, agile organization, change management*. Springer Fachmedien GmbH.

Linnhoff-Popien, C., & Schneider, R. (2018). *Digital Marketplace unleashed*. Springer Verlag GmbH.

McKinsey & Company. (2014). Connected car, automotive value chain unbound. In *Advanced industries*, S. 7–50. Retrieved June 2, 2021, from https://www.mckinsey.de/files/mck_connected_car_report.pdf

McKinsey & Company II. (2015). *Urban mobility at a tipping point*, S. 3–22. Retrieved May 6, 2021, from https://www.mckinsey.com/business-functions/sustainability-and-resource-productivity/our-insights/urban-mobility-at-a-tipping-point

McKinsey & Company I. (2016). *Automotive revolution—perspective towards 2030. How the convergence of disruptive technology-driven trends could transform the auto industry*, S. 3–19. Retrieved April 6, 2021, from https://www.mckinsey.de/files/automotive revolution_perspective_towards_2030.pdf

McKinsey & Company I. (2017). *An integrated perspective on the future of mobility, part 2: Transforming urban delivery*.

McKinsey & Company II. (2017). *The changing aftermarket game. how automotive suppliers can benefit from arising opportunities*, S. 1–36. Retrieved February 26, 2021, from https://www.mckinsey.com/~/media/McKinsey/Industries/Automotive%20and%20Assembly/Our%20Insights/The%20changing%20aftermarket%20game%20and%20how%20automotive%20suppliers%20 can%20benefit%20from%20arising%20opportunities/The-changing-aftermarket-game.ashx

McKinsey & Company III. (2016). Monetizing car data. New service business opportunities to create new customer benefits. In *Advanced industries*, S. 3–57. Retrieved February 22, 2021, from https://www.mckinsey.com/~/media/McKinsey/Industries/Automotive%20and%20Assembly/Our%20Insights/Monetizing%20car%20data/Monetizing-car-data.ashx

McKinsey & Company III. (2017). *The future(s) of mobility: How cities can benefit. Sustainability & resource productivity*, S. 1–12. Retrieved February 26, 2021, from https://www.mckinsey.com/business-functions/sustainability-and-resource-productivity/our-insights/the-futures-of-mobility-how-cities-can-benefit

McKinsey & Company V. (2016). *Urban mobility 2030: How cities can realize the economic effects. Case study Berlin*, S. 1–30. Retrieved May 6, 2021, from https://www.mckinsey.de/files/urban_mobility_english.pdf

McKinsey & Company, Bloomberg. (2016). *An integrated perspective on the future of mobility*, S. 5–62. Retrieved April 27, 2021, from https://www.bbhub.io/bnef/sites/4/2016/10/BNEF_McKinsey_The-Future-of-Mobility_11-10-16.pdf

Meier, A., & Portmann, E. (2016). *Smart city. Strategie, governance und projekte*. Springer Fachmedien GmbH.

Meyer, G., & Beiker, S. (2018). *Road vehicle automation 5. Lecture notes in mobility*. Springer International Publishing AG.

Meyer, G., & Shaheen, S. (2017). *Disrupting mobility. Impacts of sharing economy and innovative transportation on cities*. Springer International.

Morgan Stanley. (2012). *Global auto scenarios 2022. Disruption and Opportunity start now*, S. 1–45. Retrieved April 6, 2021, from https://www.morganstanleyfa.com/public/projectfiles/22e7c980-260b-4e70-9c63-8e890dec95e1.pdf

Natalia, N. (2013). *Trends in automotive industry: New mobility concept. Rethinking current business models of OEMs 2013*, S. 1–96. Retrieved June 2, 2021, from http://www.makingsciencenews.com/catalogue/papers/217/download

Neckermann, L. (2015). *The mobility revolution*. Troubador Publishing Ltd.

Nieuwenhuis, P., & Wells, P. (2015). *The global automotive industry* (1. Aufl.). John Wiley & Sons, Ltd.

OECD/International Transport Forum. (2015). *Automated and autonomous driving. Regulation under uncertainty*, S. 1–32. Retrieved April 27, 2021, from https://www.itf-oecd.org/sites/default/files/docs/15cpb_autonomous-driving.pdf

Oswald, G., & Krcmar, H. (2018). *Digitale Transformation. Fallbeispiele und Branchenanalysen.* Springer Gabler.

PA Consulting. (2017). *Autonomous vehicles—what are the roadblocks*, S.1–16. Retrieved June 6, 2021, from https://www.paconsulting.com/insights/2017/autonomous-vehicles/

Pollak, D. (2017). *Like I see it. Obstacles and opportunities shaping the future of retail automotive.* vAuto Press.

PWC. (2018). *Five trends transforming the automotive industry*, S. 1–48. Retrieved April 27, 2021, from https://www.pwc.at/de/publikationen/branchen-und-wirtschaftsstudien/eascy-five-trends-transforming-the-automotive-industry_2018.pdf

PWC - Strategy&. (2016). *Connected car report 2016. Opportunities, risk, and turmoil on the road to autonomous vehicles*, S. 5–63. Retrieved April 27, 2021, from https://www.strategyand.pwc.com/reports/connected-car-2016-study

PWC - Strategy&. (2017). *The 2017 strategy& digital auto report. Fast and furious: Why making money in the "roboconomy" is getting harder*, S. 1–41. Retrieved April 27, 2021, from https://www.strategyand.pwc.com/reports/fast-and-furious

Roland Berger. (2011). *Automotive landscape 2025: Opportunities and challenges ahead*, S. 1–47. Retrieved July 12, 2021, from http://www.forum-elektromobilitaet.ch/fileadmin/DATA_Forum/Publikationen/Roland_Berger_2011_Automotive_Landscape_2025_E_20110228.pdf

Roland Berger. (2014). Autonomous Driving. Disruptive Innovation that promises to change the automotive industry as we know it - it's time for every player to think:act! In *Think act*, S. 1–24. Retrieved March 29, 2021, from https://www.rolandberger.com/publications/publication_pdf/roland_berger_tab_autonomous_driving.pdf

Roland Berger. (2015). *Maßkonfektion im Aftersales. Servicedifferenzierung entlang der Kundenwünsche*, S. 2–51. Retrieved May 6, 2021, from http://docplayer.org/3730394-Masskonfektion-im-aftersales.html

Roland Berger II. (2016). *A CEO agenda for the (r)evolution of the automotive ecosystem. New archetypes will emerge in the future to divide the market up among themselves. How to gain access to tomorrow's profit pools*, S. 2–15. Retrieved May 6, 2021, from https://www.rolandberger.com/de/Blog/CEO-agenda-for-the-(r)evolution-of-the-automotive-ecosystem.html

Roland Berger, Lazard. (2016). *Global automotive supplier study. Being prepard for uncertainty*, S.3–41. Retrieved May 28, 2021, from https://www.rolandberger.com/publications/publication_pdf/roland_berger_global_automotive_supplier_2016_final.pdf

Rosenzweig, J., & Bartl, M. (2015). *The making of innovation. A review and analysis of literature on autonomous driving*, S. 1–56. Retrieved May 11, 2021, from http://www.michaelbartl.com/co-creation/wp-content/uploads/Lit-Review-AD_MoI.pdf

Siegfried, P. (2014). *Analysis of the service research studies in the German research field, performance measurement and management.* Publishing House of Wroclaw University of Economics, Band 345, S. 94–104.

Siegfried, P. (2015). *Trendentwicklung und strategische Ausrichtung von KMUs.* EUL-Verlag, Siegburg.

Siegfried, P., & Zhang, J. J. (2021). Developing a sustainable concept for the urban lastmile delivery. *Open Journal of Business and Management (OJBM).* https://doi.org/10.4236/ojbm.2021.91015

Simpson, T., Siddique, Z., & Jiao, J. (2006). *Product platform and product family design. Methods and applications.* Springer Science + Media Business, Inc.

Skilton, M. (2016). *Building digital ecosystem architectures. A guide to enterprise architecting digital technologies in the digital enterprise* (1. Aufl.). Palgrave Macmillan.

The Boston Consulting Group. (2004). *Beyond cost reduction. Reinventing the automotive OEM-supplier interface*, S. 3–48. Retrieved April 23, 2021, from https://www.bcgperspectives.com/content/articles/automotive_sourcing_procurement_beyond_cost_reduction_reinventing_automotive_oem_supplier_interface/

The Boston Consulting Group, World Economic Forum. (2016). *Self-driving vehicles, robot-taxis, and the urban mobility revolution*, S. 3–26. Retrieved March 29, 2021, from https://www.bcg.com/…/automotive-public-sector-self-driving-vehicles-robo-taxis-urban-mobility-revolution.aspx

U.S. Department of Commerce, Economics and Statistics Administration. (2017). The employment impact of autonomous vehicles. In *ESA Issue Brief* (5), S. 1–33. Retrieved March 12, 2021, from http://www.esa.doc.gov/sites/default/files/Employment-Impact Autonomous Vehicles_0.pdf

UBS. (2017). *Longer term investments. Smart mobility*, S. 1–25. Retrieved March 12, 2021, from https://www.ubs.com/content/dam/WealthManagement Americas/documents/smart-mobility.pdf

Vo, P. H. (2015). *Die Automobilindustrie und die Bedeutung innovativer Industrie 4.0 Technologien.* Diplomica Verlag GmbH.

Vogel, H.-J., Weißer, K., & Hartmann, W. D. (2018). *Smart city: Digitalisierung in Stadt und Land. Herausforderungen und Handlungsfelder.* Springer Fachmedien GmbH.

Waschl, H., & Kolmanovsky, I. (2018). *Control strategies for advanced driver assistance systems and autonomous driving functions. Development, testing and verification*. Springer International Publishing AG.

Wedeniwski, S. (2015). *The Mobility Revolution in the Automotive Industry. How not to miss the digital turnpike*. Springer Verlag GmbH.

Winkelhake, U. (2017). *Die digitale Transformation der Automobilindustrie. Treiber - Roadmap - Praxis*. Springer Verlag GmbH.

Wyman, Oliver IV (2016). Sourcing in the automotive industry: How can suppliers create more value? In *Insights on automotive supplier excellence*, S. 2–10. Retrieved April 23, 2021, from http://www.oliverwyman.com/content/dam/oliver-wyman/v2/publications/2016/jan/OliverWyman_Sourcing_in_the_AutomotiveIndustry_web.pdf

6

Conclusion to Automotive Disruption and the Urban Mobility Revolution

6.1 Outlook 2030

The subsequent wrap-up will briefly abstract the most significant findings of this research work and review the introductory questions that have remained unanswered.

How will the transformation progress and what are the key drivers? The upheaval of the automotive industry is a drawn-out process induced by a variety of economic, societal, technological, and political key drivers. Ethical consumerism and the legislative push for ecological sustainability have set the breeding ground for the electrification of the car and the search for alternative mobility models. The customer pull for mobility solutions that are adaptable to individual lifestyles has created an incentive for the automotive industry and service providers to commercialize and monetize novel mobility models and related digital features. Customer- and vehicle-related data has emerged as a strategic asset to deliver additional value to consumers by tailoring product and service offerings according to personal preferences. The increasing integration of data into the mobility experience will simultaneously proceed with the advancements in autonomous driving technology. The pace of the transformation and its final phenotype will be highly dependent on the interaction of those two determining factors. A major share of the economic potential is ascribed to the usage of the vehicle as a third space in which automotive manufacturers can generate revenues by introducing automated driving features and service providers can commercialize the idle time of passengers by providing various digital applications via the human-machine-interfaces.[1] The human factor will thus be the most decisive criterion for the success of the new mobility experience. Customers' acceptance and usage of automotive innovations are the prerequisites for their economic and financial viability.

Will the traditional business model of OEMs cease to exist? The new focus on service-centricity does not abrogate the need for vehicles in the long term. The incremental shift from internal combustion engines towards alternative powertrains and electrified cars will, in fact, lead to a reorganization of the product portfolio causing a restructuring of global value chains,[2] but the basic principles of automobile production will remain untouched. However, the diffusion of EVs and on-demand and shared mobility models will be a successive, fragmented process.

While industrialized nations represent an incubator for these new vehicle technologies and business models, developing and emerging countries will further depend on traditional forms of locomotion. This gives automotive manufacturers the opportunity to parallelly market both products, the classic and the novel mobility. In conclusion, the diffusion of mobility innovations will not replace traditional mobility but increase its diversity.

How will the automotive landscape look like in 2030? The automotive industry is facing its biggest upheaval since its origination. The speed and dimension of the transformation make a distinct outlook impossible and the multiplicity of internal and external determining parameters cannot be taken into account in their entirety. Even economics literature and professional articles show a significant divergence in the evaluation and illustration of future scenarios.[3] As this research work has shown, the

---

[1] cf. Blanke, T. (2014) Digital asset ecosystems S.21.
[2] cf. PwC (2017) PwC automotive industry bluebook (2017 Edition) S.9–13.
[3] cf. McKinsey & Company I (2016) Automotive revolution—Perspective towards 2030 S.12.

automotive revolution can rather be considered an evolution, since the transformational process will not be completed in 2030. Nobody can prophesy with absolute certainty how the automotive industry will evolve when leaving its larvae stage.

6.2 Limitations of the Research Work

The magnitude of transformation that is altering the automotive industry in the recent decade exceeds the dimension of change that could be observed in the last century. Innovations impacting the concept of mobility are introduced at a breathtaking pace. The digital era has generated a new level of innovative capacity that makes it nearly impossible to anticipate technological progress for the upcoming years. In consequence, this research work should be considered a snapshot of the current level of development. A portrait of the automotive landscape in its entirety in 2030 will always have a hypothetical character since an accurate determination of the prospective path of automotive (r)evolution is always constrained by assumptions. This also applies to the survey conducted within the framework of this research work and the studies referred to, to complement the conjectures and key results. The limited number of survey participants impedes the provision of a representative image on a societal level. As indicated in Sect. 4.1, the survey results might be biased based on respondents' demographic peculiarities that deviate from the average values.[4] The determination of the prospective business volume and penetration levels are thus to be handled with care.

Although the external studies, cited in this research work, cover by far larger survey groups, many parameters used within the calculation of future valuation models are based on assumptions and hypotheses. This applies, in particular, to the penetration of smart mobility services, the identification and quantitation of revenue pools and consumers' prospective spending behavior. Legislative and political influencing factors aren't

---

[4] cf. KPMG (2015) The clockspeed dilemma S,7ff.

attached sufficient importance or are partially disregarded.[5] However, the enacting of German and European laws that regulate data rights and protection, testing and licensing of autonomous vehicles or liability of insurance claims is a decisive criterion in appraising the pace of technological development and market penetration. The pioneering position of the German automotive industry is jeopardized by predetermined market conditions. The protectionist measures erected by the Chinese government in combination with extensive subsidies granted to promote advanced mobility models and related digital services might lead to a dislocation of the incubator of the automotive disruption.[6] Consequently, the commercialization and market introduction of the aforementioned innovations could be delayed, voiding the predicted scenarios.

In conclusion, this research work reflects an image of the Business Model 2030 based on the current status quo of the automotive industry. Prospective developments are subject to a multitude of determining factors that cannot be considered in their entirety at this juncture.

References

Blanke, T. (2014). *Digital asset ecosystems. Rethinking crowds and clouds*. Chandos Publishing.

Deloitte University Press. (2015). *Patterns of disruption. Anticipating disruptive strategies in a world of unicorns, black swans and exponentials*, S. 1–34. Retrieved May 11, 2021, from https://www2.deloitte.com/content/dam/Deloitte/br/Documents/technology/Patterns-of-disruption.pdf

KPMG. (2015). *The clockspeed dilemma. What does it mean for automotve innovation?* S. 1–40. Retrieved February 26, 2021, from https://assets.kpmg.com/content/dam/kpmg/pdf/2016/04/auto-clockspeed-dilemma.pdf

McKinsey & Company I. (2016). *Automotive revolution—perspective towards 2030. How the convergence of disruptive technology-driven trends could transform the auto industry*, S. 3–19. Retrieved April 6, 2021, from https://www.mckinsey.de/files/automotive_revolution_perspective_towards_2030.pdf

---

[5] cf. Deloitte University Press (2015) Patterns of disruption S.14–17.
[6] cf. Roland Berger; fka Forschungsgesellschaft (2016) Index "Automatisierte Fahrzeuge".

PWC. (2017). *PwC automotive industry bluebook (2017 edition) China automotive market. Witnessing the transformation*, S. 1–38. Retrieved April 6, 2021, from https://www.pwccn.com/en/automotive/pwc-auto-industry-blue-book.pdf

Roland Berger; fka Forschungsgesellschaft. (2016). *Index "Automatisierte Fahrzeuge"*, S. 2–18. Retrieved May 11, 2021, from https://www.roland-berger.com/publications/publication_pdf/roland_berger_index_autonomous_driving_q3_2016_final_d.pdf

7

Prospects: A Look Beyond

7.1 Future Aspects

For a long time, autonomous driving has only been a distant vision of engineers reaching for the stars. But increasing technology maturity has paved the way for automobile producers to achieve ever-higher degrees of automation and the next level is already within their grasp. With autonomy becoming a more realistic endeavor in the mid-term, many critical voices have been raised that publicly denounce the ethical implications of the machine age. The subsequent section will give a look beyond the mere technical side and illuminate the alleged shady sides of the current developments.

In addition to computer-based simulation runs and trial runs in enclosed testing facilities, an increasing number of IT companies and start-ups developing autonomous vehicles are receiving authorization for testing prototypes in public traffic under realistic conditions. In March 2018, the first pedestrian death associated with self-driving technology has occurred when an autonomous Uber fatally hit a woman while crossing the street.[1] This incident has caused a re-ignition of the global

---
[1] cf. The New York Times (2018) How a self-driving Uber killed a pedestrian in Arizona.

debate about SDV's safety degree. While roughly 35,000 traffic participants have died in lethal accidents in the USA in 2016, thereof 86% caused by human error, autonomous vehicles are apparently subject to significantly stricter safety requirements.[2] The objective of zero-fatalities that is pursued by the automotive industry, however, will not be accomplished without critical incidents. Critical voices need to consider that novel innovations always suffer setbacks but in a long-term perspective, autonomous technology offers a huge potential to drastically reduce human-caused traffic injuries and deaths. Another controversy about SDVs is of ethical nature. In theory, an autonomous vehicle could recognize the age of bicyclists and pedestrians and could choose who is rather deserving protection in case of an accident. It relates to the old question if a young life is more valuable than an old life. Ethics commissions are still debating how to configure the emergency program of SDVs to resolve this dilemma by possibly using random generators that do not consider the age and physical constitution or by computing the degree of harm inflicted to all traffic participants affected by the accident. The fundamental idea of these approaches is that the value of human life is not assessable.

Another major concern is that the theory of the "transparent citizen" could assume a definite form. Since connectivity features and digital services access and store personal and vehicle-related data, critical voices already propagandize a surveillance of the users by private business and governmental institutions. The newly installed Chinese social point system is often cited as a deterrent example for how the omnipresent availability of personal information could backfire.[3] Personal data could be analyzed to determine contraventions, traffic misconduct, privacy-sensitive preferences, or daily routines. Politics and legal authorities need to take action to establish a democratically based environment that sets respective regulatory frameworks and protects civic rights.

In conclusion, the automotive industry is a continuous construction site or—applying the semantics used in the introduction—a vivid organism that passes through the next development stage.

---

[2] cf. OECD; ITF (2018) ITF transport statistics. Road accidents.
[3] cf. Manager Magazin (2018) Punktabzug für Autofahrer und Demonstranten.

References

Manager Magazin. (2018). *Punktabzug für Autofahrer und Demonstranten—so bewertet China seine Bürger. Punktesystem als Mittel totaler sozialer Kontrolle.* Retrieved March 5, 2021, from http://www.manager-magazin.de/politik/artikel/china-soziale-kontrolledurch-sozialkredit-punktesystem-a-1194590.html

OECD; ITF. (2018). *ITF transport statistics. Road accidents.* Retrieved July 15, 2021, from https://www.oecd-ilibrary.org/transport/data/itf-transport-statistics/road-accidents_g2g55585-en

The New York Times. (2018). *How a self-driving Uber killed a pedestrian in Arizona. Unter Mitarbeit von Troy Griggs und Daisuke Wakabayashi.* Retrieved July 17, 2021, from https://www.nytimes.com/interactive/2018/03/20/us/self-driving-uber-pedestrian-killed.html